CONTROL SYSTEMS

MODELING, ANALYSIS, AND DESIGN

FIRST EDITION

Swami Karunamoorthy
WASHINGTON UNIVERSITY, ST. LOUIS

cognella®

SAN DIEGO

Bassim Hamadeh, CEO and Publisher
Angela Schultz, Senior Field Acquisitions Editor
Albert Liau, Project Editor
Susana Christie, Senior Developmental Editor
Jordan Krikorian, Editorial Assistant
Celeste Paed, Associate Production Editor
Emely Villavicencio, Senior Graphic Designer
Alexa Lucido, Licensing Manager
Bonnie Dietrich, Interior Designer
Stephanie Adams, Senior Marketing Program Manager
Natalie Piccotti, Director of Marketing
Kassie Graves, Senior Vice President, Editorial
Jamie Giganti, Director of Academic Publishing

Printed in the United States of America.

cognella® | ACADEMIC PUBLISHING
3970 Sorrento Valley Blvd., Ste. 500, San Diego, CA 92121

BRIEF CONTENTS

DETAILED CONTENTS

CHAPTER 3

Modeling of Control Systems 31

CHAPTER 4

Characteristics of Control Systems 56

PREFACE

In a physical system, if the desired output is used as input, the measured actual output should match with the input. If there is an error in this process, there is a need for a control system to rectify the error so that the output matches with the input. Examples include flushing a toilet; heating or cooling the house; using cruise control in a car; controlling an airplane, a wind turbine, or a space station; managing blood pressure during surgery—these are all real systems and require knowledge of their control. So, it is necessary to learn this topic in the undergraduate curriculum in mechanical, biomedical, aerospace, chemical, electrical, mechatronics, and similar branches of engineering and technology. This book satisfies the academic merit in all of the above areas.

This book's emphasis is on mechanical engineering systems and provides lessons that can easily be applied to aerospace, biomedical, and mechatronics systems. The examples and case studies offered in the abovementioned areas give this book a unique perspective.

The approach this book uses is based on a dynamic mechanical system. The pedagogical approach entails learning first the modeling, analysis, and design of negative-feedback systems and then those of state variable modern control systems. The book also discusses the use of MATLAB and Simulink, the modern engineering tools, and provides the required knowledge to satisfy some of the Accreditation Board for Engineering and Technology's student outcomes. These are criterion 3-(1) to solve complex engineering problems; criterion 3-(2) to design with respect to the components of control system; and criterion 5-(b) to utilize modern engineering tools.

This book serves as a foundation or prerequisite for advanced courses in nonlinear control systems, digital control systems, aircraft stability and control, space vehicle control, wind turbine control, rotary wing control, biomedical control systems, robotics control, adaptive control, and chemical process control.

INTRODUCTION

This book serves as a resource to undergraduate students in both engineering and technology in the areas of mechanical, aerospace, biomedical, mechatronics, and electrical. Students are assumed to have previous experience with dynamics, differential equations, Laplace transform, and matrices. These tools are applied in the modeling of dynamic and control systems and the study of the characteristics, performance, stability, response, and design of several control systems. A state variable model is also presented to provide an introduction to modern control systems.

The organization of the book's chapters is similar to that of a typical control system diagram, as shown below. Learning a course can also be perceived as a control system. A typical chapter begins with an introduction to describe what we will learn in the chapter. Next, the learning objectives serve to display the desired knowledge used as input to the learning system. Topic discussion within the chapters serves as pedagogical education to attain the desired knowledge. At the end of each chapter, a summary is provided to describe what was taught within it. The acquired new knowledge after reading each chapter should match with desired knowledge given at its beginning. Following the summary, multiple-choice questions and practice problems are included as an assessment to demonstrate accomplishment of the learning objectives. The test also provides feedback to verify that the knowledge learned is consistent with that which was desired, as shown in the control system diagram. The process is repeated in each chapter for effective learning.

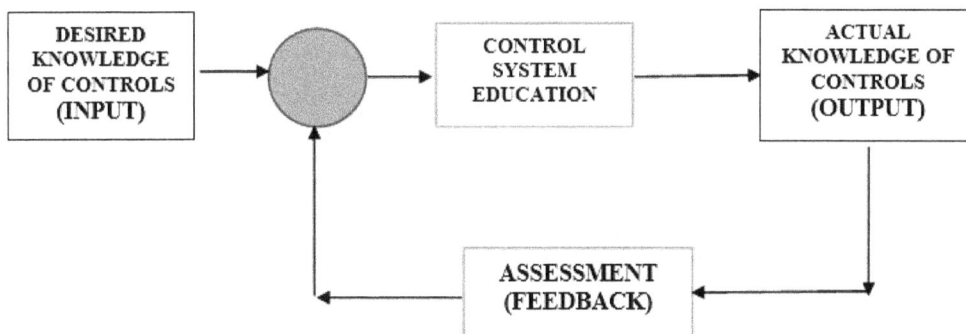

Image 0.1 Control System Diagram

The topics are presented using a student-centered approach for easy understanding and effective learning. In each chapter, several examples are given to learn the application of theory. Some of these examples are specific to mechanical, aerospace, or biomedical systems. MATLAB is used as a tool for complex numerical problems. The commonly used topics in MATLAB are given in Appendix A, including a table with a list of typical commands. Simulink is an important tool for the simulation of a control system and has graphical output in order to visualize a system's behavior. Appendix A also includes a tutorial on how to use the tool as well as an example. The application of both MATLAB and Simulink to typical problems in a control system is included in the practice problems within chapters 2, 3, and 7. The answers to these problems are provided in Appendix B.

CONSTITUENTS OF CONTROL SYSTEMS

1.1. Introduction

This chapter introduces the constituents or building blocks of a typical control system. It would help to understand what a control system is and its components. A control system is a study of methods for modeling, analysis, and design to accomplish an output that is equal to or very close to the desired output that is used as the system's input.

1.2. Learning Objectives

1. Understand the various types of control systems.

2. Identify the various elements of a control system.

3. Learn about transfer functions.

1.3. Types of Control Systems

The types of control system are generally classified as (i) open-loop and (ii) closed-loop. The open-loop control system has no feedback and does not depend on output. Because the information is forwarded from start to finish, it is also called a *feed-forward system*. In a closed-loop system, sensors are used to measure the output and the feedback to the input that alters the desired output, until the output matches the input. This type of system is also called a *feedback control system*.

1.3.1. Open-Loop (or Feed-Forward) Systems

Microwave ovens, bread toasters, and space heaters are typical devices with an open-loop control system.

These systems function as a typical on-off switch. They do not control the temperature to which food should be heated in a microwave, how well bread should be toasted, or how warm a room should be heated.

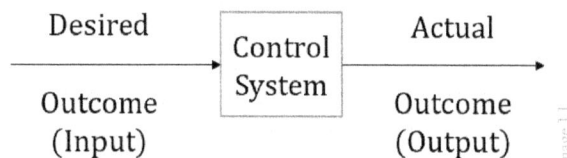

Desired | Control System | Actual

Outcome (Input)   Outcome (Output)

Image 1.1

1.3.2. Closed-Loop (or Feedback) Systems

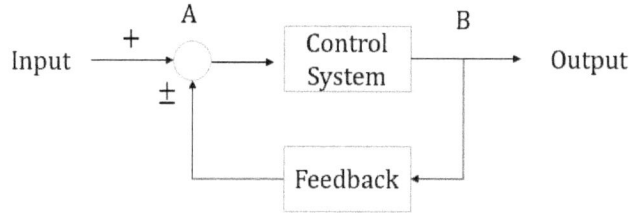

A. Summing point

B. Pick-off point

Temperature-controlled ovens, items heated using a thermostat, and a car's cruise control system are typical examples of closed-loop systems. The heating turns on or off automatically when the actual heat is below or above the desired heat level in order to maintain the set level of heat in both an oven and a house. In a car, the speed is controlled by cruise control so that the car will travel at a certain speed. Cruise control controls the fuel's flow into the engine. When a car is traveling uphill, its speed drops, and its cruise control increases the fuel output (equivalent to a driver pushing on the accelerator) into the engine. When a car is going downhill, its speed increases, and its cruise control decreases the fuel flow to the engine to maintain a set level of speed.

Types of feedback can be further classified as positive and negative feedback.

1.3.3. Positive Feedback

Let $\Delta =$ (desired output $-$ actual output)

The value of Δ is used as feedback. If its value decreases, the actual output converges closer to the desired output. The decreased value of Δ is attributed to negative feedback, whereas the increased value of Δ is attributed to positive feedback. The feedback from actual output is added to the input, making it larger in positive feedback.

For example, if a person speaks through a microphone, their voice is amplified to a certain level through the speaker. If the microphone is close to the speaker,

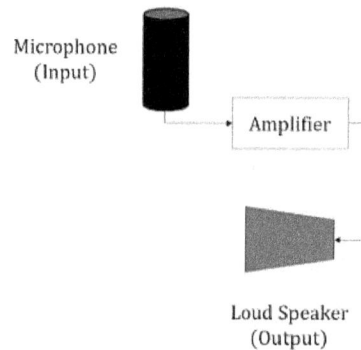

Image 1.3   Public Address System

the output from the speaker is fed continuously into the microphone, thus increasing the desired output to a high level. This positive feedback results in a loud squealing or howling noise. Therefore, positive feedback makes the input larger instead of minimizing the difference between the input and output. Thus, in general, positive feedback leads to system instability and is undesirable in the control of engineering systems.

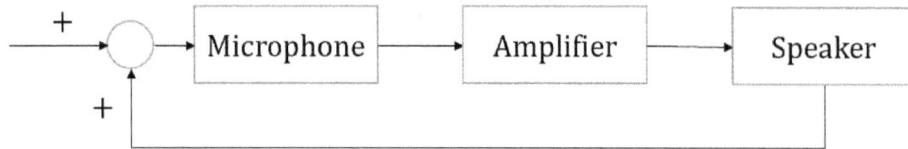

Image 1.4   Control Diagram

1.3.4. Negative Feedback

In a negative-feedback system, the feedback from the actual output is subtracted from the input to control the difference between them.

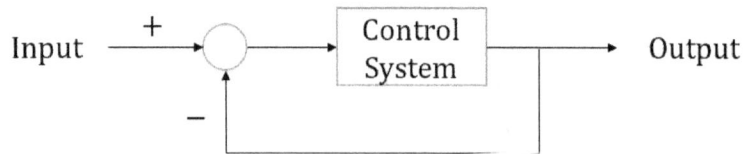

For example, in a home's centralized heating system, the thermostat controls the temperature to a set value.

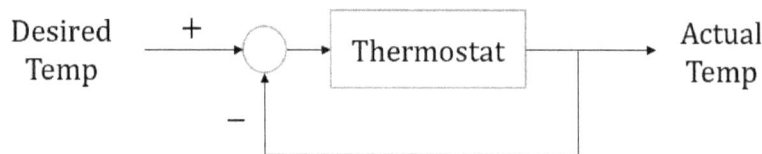

If the actual temperature is less than that which is desired, the feedback to the thermostat activates the heating system to increase the house's temperature. It turns off the heating system when the feedback indicates that the actual temperature is greater than or equal to the desired temperature. Therefore, negative feedback systems lead to system stability and hence are commonly used in control of engineering systems.

Emphasis will be placed here on understanding, modeling, analyzing, and designing negative feedback automatic control systems, which are also called *modern control systems*.

1.4. Elements of Control Systems

The various elements of a typical control system are as follows:

    i. Controller

  ii. Actuator

 iii. Plant or process

 iv. Sensor

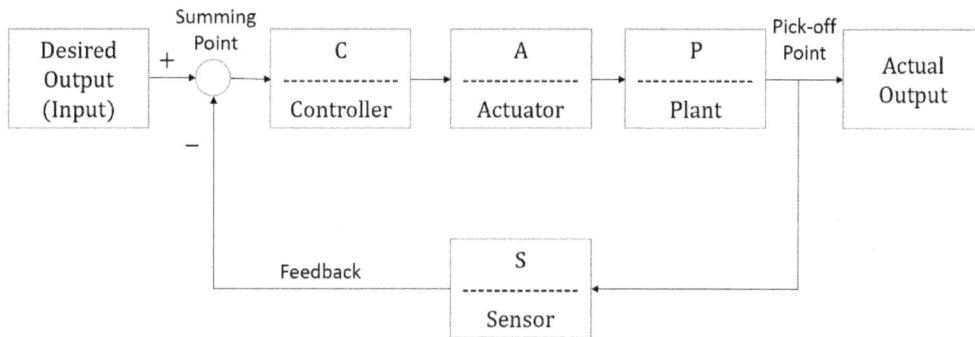

The control diagram above represents a closed-loop system, and without the feedback loop, it becomes an open-loop system, as shown below.

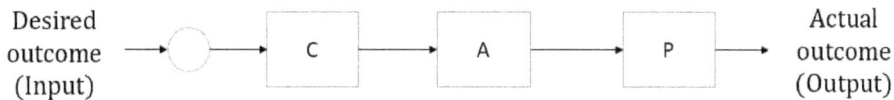

If the actual output is equal or very close to the desired output, an open-loop system is adequate for use. When the output is not equal to that which is desired, it is not only necessary but also required to use a closed-loop system with negative feedback. This ensures both system control and stability.

Let us look at a typical home's heating control system and its control elements.

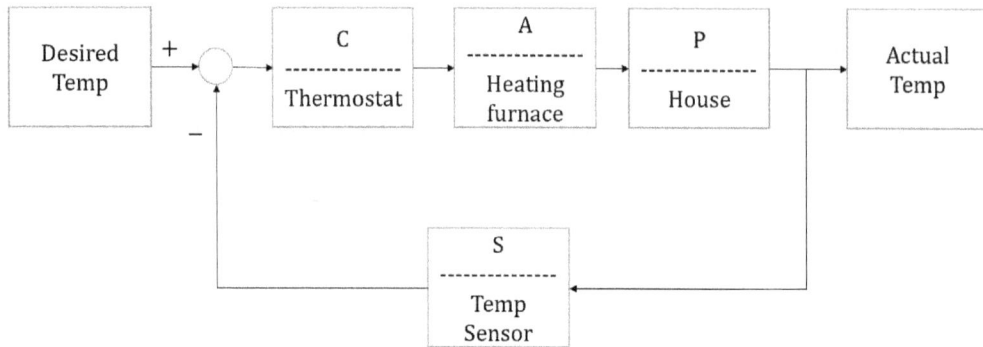

Controller: device used to activate the actuator. In the above example, a thermostat is used as a controller to turn on or off the heating furnace. If a person were used instead for this operation, this would be manual control. Because a device is used, it is automatic control.

Actuator: device in the control system used to alter or adjust the environment. In the above example, the heating furnace is the actuator and adjusts the home's heat by switching it on or off.

Plant: system (device, industrial plant, or process) under control. In the above example, the house is the plant, and its environment or temperature is being controlled to a set level of comfort.

Sensor: device that provides the measurement of the actual output. In the above example, the temperature-sensing device within the thermostat is the sensor element in the control system.

A typical temperature-sensing device is a bimetallic strip. It is used to make or break contact with an electric circuit, which turns on or off the heating system.

1.5. Cruise Control in a Car

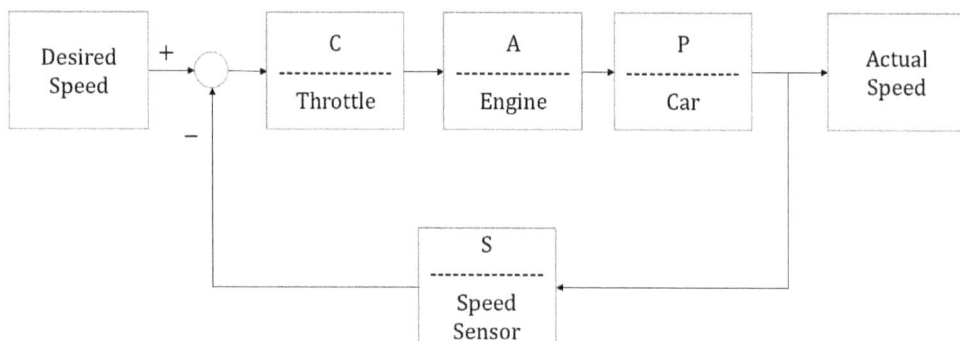

Another example is a car's cruise control. Here, the controller is a throttle that regulates the fuel flow.

The engine is the activator that increases or decreases the car's speed. The car is the plant being controlled in order to travel at a certain speed. The speed sensor gives feedback on any change to the actual speed with respect to the desired speed. Based on this feedback, the throttle will adjust the fuel flow into the engine.

1.6. Learning Process Control

A student's learning process can be represented as a negative-feedback control system.

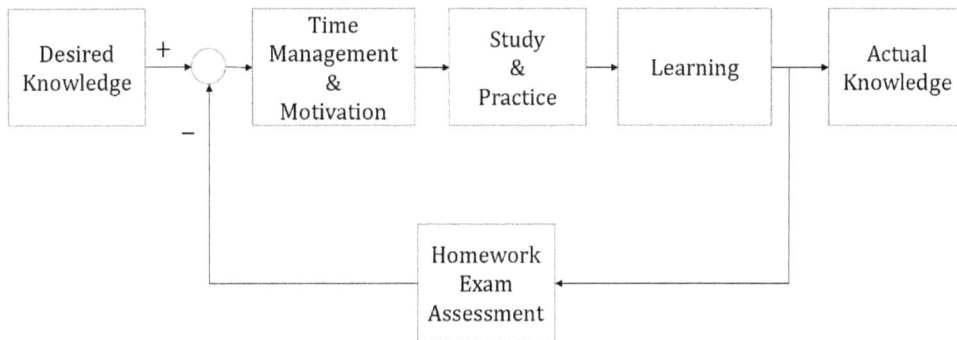

In classroom learning, the course outline sets the goal for desired knowledge. Motivation and time management are the controllers in this process; studying the book or class notes and practicing problems are the actuators. The plant in this system is the learning process that is controlled to gain the needed knowledge. Homework, examinations, and assessment of course outcomes are the sensors used to measure the actual knowledge gained and are used as feedback for better learning.

1.7. Transfer Function

Transfer function (TF) is widely used in the study of control systems. This function is defined as the ratio of output to input in the Laplace transform (LT) domain, with all initial conditions assumed to be zero.

1.7.1. Open Loop

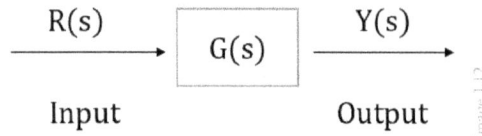

Image 1.12

$$Y(s) = G(s)R(s)$$

$$G(s) = \frac{Y(s)}{R(s)} = \frac{Output}{Input}$$

Where

S = frequency domain in LT

$R(s)$ = LT of input

$Y(s)$ = LT of output

$G(s)$ = TF

1.7.2. Closed Loop

NEGATIVE FEEDBACK

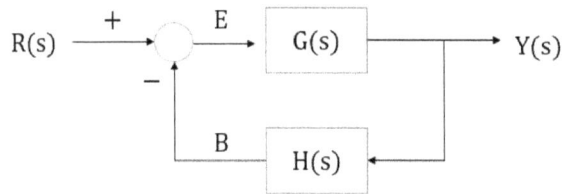

Image 1.13

Where:

$G(s)$ = feed-forward TF

$H(s)$ = feedback TF

E/R = error ratio

Y/R = control ratio

B/R = feedback ratio = $\begin{cases} \text{Negative} \rightarrow \text{Negative feedback} \\ \text{Positive} \rightarrow \text{Positive feedback} \end{cases}$

$E = R\text{-}B$

$G = Y/E$

$H = B/Y$

$Y = GE = G(R - B) = G(R - HY) = GR - GHY$

$Y(1 + GH) = GR$

TF:

$Y/R = G/(1 + GH) = $ output/input

POSITIVE FEEDBACK

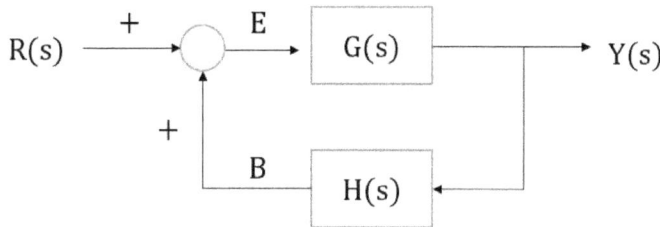

$E = R + B \, ; B = HY$

$Y = GE = G(R + B) = GR + GHY$

$Y(1 - GH) = GR$

TF:

$Y/R = G/(1 - GH) = $ output/input

If $GH \cong 1$, the output may become infinite, making the system unstable.

1.8. Summary

The open-loop and closed-loop control systems are primary constituents of a typical control system. The closed-loop system consists of a positive-feedback system and a negative-feedback system. The negative-feedback system is commonly encountered in engineering applications. The building blocks of a negative-feedback control system are the controller, actuator, plant (or process), and sensor. These elements are identified in a thermostat, cruise control, and learning process in education, for example.

The TF is the key parameter to study a control system, and it has been defined and also derived for both open-loop and closed-loop systems. The closed-loop TF includes both negative-feedback and positive-feedback systems. It has been shown that positive feedback makes a system unstable and hence it is not commonly used in engineering applications.

1.9. Assessment

Answer the following multiple-choice concept questions:

1. Control system is a study to control:

   a. The failure of a system

   b. The operation of a function

   c. The actual output to match the desired output

   d. Control the input

2. An open-loop control system:

   a. Depends on output

   b. Depends on input

   c. Does not depend on input

   d. Does not depend on output

3. A closed-loop control system:

   a. Depends on output

   b. Provides feedback

   c. Provides stability

   d. All of the above

4. A controller is defined as:

   a. A device that activates the plant

   b. A device that activates the actuator

   c. A device that controls the sensor

   d. All of the above

5. An actuator is defined as:

   a. A device that activates the plant

   b. A device that activates the controller

   c. A device that activates the sensor

   d. All of the above

6. A plant or process is defined as a system:

   a. That controls the actuator

   b. That controls the sensor

   c. That activates the controller

   d. To be controlled

7. A sensor is defined as:

   a. A device to measure the output from a plant

   b. A device to give feedback to the controller

   c. A device to function as a sensing element in the system

   d. All of the above

8. Transfer function is defined as the:

   a. Ratio of input to output

   b. Ratio of output to input

   c. Sum of output and input

   d. Difference between output and input

9. $G/(1 + GH)$ is the transfer function for:

   a. An open-loop system

   b. A positive-feedback system

   c. A negative-feedback system

   d. None of the above

10. $G/(1 - GH)$ is the transfer function for:

    a. An open-loop system

    b. A positive-feedback system

    c. A negative-feedback system

    d. None of the above

1.10. Practice Problems

1. Answer all the assessment questions in Section 1.9.

2. Sketch and describe the block diagram of the control system for a toaster as an open-loop system (use online resources).

3. Sketch and describe the block diagram of the control system for a toaster as a closed-loop system (use online resources).

4. Identify various sensors used in a thermostat for a climate control system and briefly describe them (use online resources).

5. Identify five different sensors and their functions used in a human system.

6. Briefly describe the control system in an autonomous car (use online resources).

7. Given the input $R(t)$ = unit step function $[U(t)]$ and the output $y(t) = e^{-at}$, determine the transfer function.

8. A linear spring stretches x inches when a force F pounds is applied. Draw a block diagram to represent this process with the force as input and the deformation (or stretch) as output.

9. Determine the transfer function for the negative-feedback system given below:

$$G(s) = (1/s); H(s) = 1/(s + 4)$$

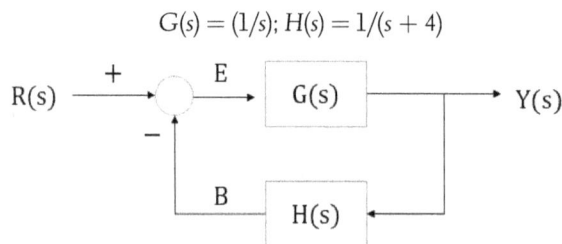

10. Determine the transfer function for the positive-feedback system given below:

$$G(s) = 1/(s^2); H(s) = s(s - 4)/(s + 6)$$

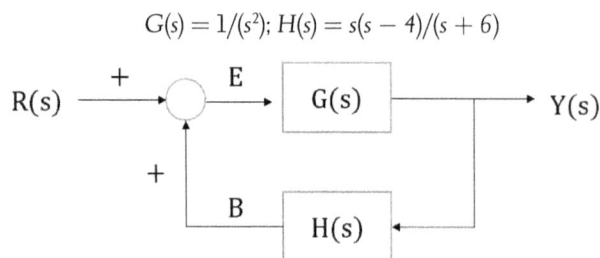

2

MODELING OF DYNAMIC SYSTEMS

2.1. Introduction

The study of control systems is based on the transfer function (TF) of a given physical system. This chapter describes the method to find the TF of a dynamic system with single or multiple degrees of freedom. For a single degree of freedom, the Laplace transform (LT) can be easily applied to the equation of the motion in a time domain to find the ratio of output to input in a frequency domain, which is defined as the TF. For multiple degrees of freedom, it is necessary to represent the equations of motion in a matrix format. A rule-based method is presented in this chapter to develop these in this format directly from the given dynamic system without any free-body diagram or equilibrium equations.

2.2. Learning Objectives

1. Understand the TF for a dynamic system with a single degree or multiple degrees of freedom.

2. Model the given dynamic system with equations of motion in a matrix format.

3. Understand poles, zeros, and the final value theorem.

2.3. Transfer Function for Dynamic Systems

TF, $G(s)$ = Output/Input = $Y(s)/R(s)$; where $Y(s)$ is the output and $R(s)$ is the input in the Laplace (or frequency) domain.

2.3.1. Systems with a Single Degree of Freedom

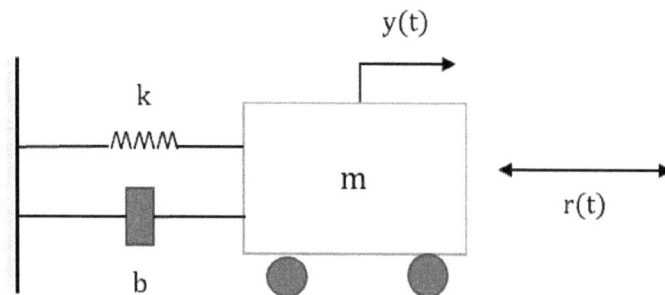

$$m = \text{mass}$$
$$b = \text{damping coefficient}$$
$$k = \text{stiffness}$$
$$y(t) = \text{output (response); } r(t) = \text{input (forcing function)}$$

Equation of motion:

$$m\ddot{y} + b\dot{y} + ky = r(t)$$

Assume the initial conditions are zero because the TF is not a function of them.

Taking the LT:

$$(ms^2 + bs + k)Y(s) = R(s)$$

$$Z(s)Y(s) = R(s); \ Z(s) = ms^2 + bs + k$$

$$Y(s) = \left[\frac{1}{Z(s)}\right] R(s)$$

$$= G(s) \, R(s)$$

Where the TF:

$$G(s) = \frac{1}{Z(s)} = \frac{1}{ms^2 + bs + k}$$

Multiple degrees of freedom:

$$[M]\{\ddot{y}\} + [C]\{\dot{y}\} + [k]\{y\} = \{r(t)\}$$

$$[M] = \text{mass matrix}$$
$$[C] = \text{damping matrix}$$
$$[K] = \text{stiffness matrix}$$

Taking the Laplace Transform:

$$\left\langle s^2[M] + s[C] + [K] \right\rangle \{Y(s)\} = \{R(s)\}$$

$$[Z(s)] \{Y(s)\} = \{R(s)\}$$

$$\{Y(s)\} = \left[Z(s)\right]^{-1}\{R(s)\}$$

$$= [G(s)] \{R(s)\}$$

Where the TF matrix:

$$[G(s)] = [Z(s)]^{-1} = \left\langle S^2[M] + S[C] + [K] \right\rangle^{-1}$$

2.3.2. Two Degrees of Freedom

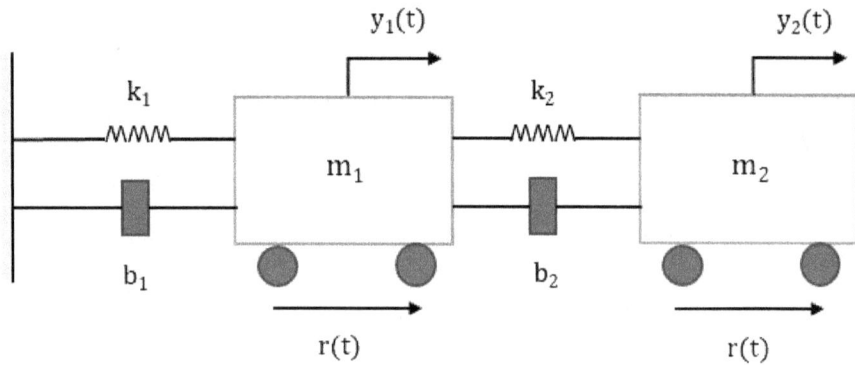

$$\left\{ \begin{array}{c} Y_1(s) \\ Y_2(s) \end{array} \right\} = \left[\begin{array}{cc} G_{11} & G_{12} \\ G_{21} & G_{22} \end{array} \right] \left\{ \begin{array}{c} R_1(s) \\ R_2(s) \end{array} \right\}$$

Let $R_2 = 0$:

$$Y_1(s) = G_{11}(s)\, R_1(s); \; G_{11}(s) = \frac{Y_1(s)}{R_1(S)}$$

$$Y_2(s) = G_{21}(s)\, R_1(s); \; G_{21}(s) = \frac{Y_2(s)}{R_1(S)}$$

Let $R_1 = 0$:

$$Y_1(s) = G_{12}(s)\, R_2(s); \; G_{12}(s) = \frac{Y_1(s)}{R_2(S)}$$

$$Y_2(s) = G_{22}(s)\, R_2(s); \; G_{22}(s) = \frac{Y_2(s)}{R_2(S)}$$

In general,

$$G_{ij}(s) = \frac{Y_i(S)}{R_j(S)}$$

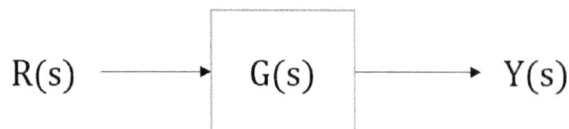

2.4. Modeling of Dynamic Systems

When a system with multiple degrees of freedom is dynamically uncoupled, the rule-based method can be applied to obtain the mass, damping, and stiffness matrices.

2.4.1. Rule-Based Method

RULE 1: DEGREE OF FREEDOM, (N)

Each mass has a degree of freedom. If there is a degree of freedom not associated with a mass, assume a virtual mass (or dummy mass) with zero value for that degree of freedom.

It helps to apply this method. The size of matrices $[M]$, $[C]$, and $[K]$ are $N \times N$.

RULE 2: MASS MATRIX, $[M]$

Because the system is assumed to be dynamically uncoupled, the mass matrix is always diagonal.

RULE 3: STIFFNESS MATRIX

Diagonal elements:
$$K_{ii} = \sum_{i=1}^{N}(\text{stiffnesses connected to Mass, } M_i)$$

Off-diagonal elements:
$$K_{ij} = -\sum_{i=1}^{N}(\text{stiffnesses connected between } M_i \text{ and } M_j)$$

$$K_{ji} = K_{ij} \text{ due to symmetry}$$

RULE 4: DAMPING MATRIX

Diagonal elements:
$$C_{ii} = \sum_{i=1}^{N}(\text{Damping coefficients connected to Mass, } M_i)$$

Off-damping elements:

$$C_{ij} = -\sum_{i=1}^{N}(\text{Damping coefficients connected between } M_i \text{ and } M_j)$$

$$C_{ji} = C_{ij} \text{ due to symmetry}$$

2.4.2. Examples

EXAMPLE 1

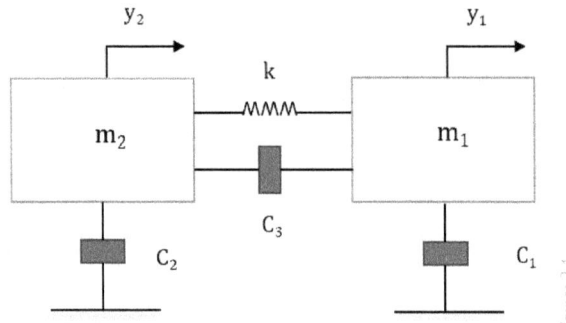

Find the TF:

$$[M] = \begin{bmatrix} m_1 & 0 \\ 0 & m_2 \end{bmatrix} \qquad [C] = \begin{bmatrix} c_1 + c_3 & -c_3 \\ -c_3 & c_2 + c_3 \end{bmatrix} \qquad [K] = \begin{bmatrix} k & -k \\ -k & k \end{bmatrix}$$

Let $m_1 = 1\,\text{kg}$, $m_2 = 2\,\text{kg}$, $c_1 = 1\,\text{kg/s}$, $c_2 = 2\,\text{kg/s}$, $c_3 = 3\,\text{kg/s}$, $k = 1\,\text{N/m}$:

$$[M] = \begin{bmatrix} 1 & 0 \\ 0 & 2 \end{bmatrix} \qquad [C] = \begin{bmatrix} 4 & -3 \\ -3 & 5 \end{bmatrix} \qquad [K] = \begin{bmatrix} 1 & -1 \\ -1 & 1 \end{bmatrix}$$

$$Z(s) = S^2[M] + S[C] + [K] = \begin{bmatrix} (s^2 + 4s + 1) & -(3s + 1) \\ -(3s + 1) & (2s^2 + 5s + 1) \end{bmatrix}$$

Transfer Function

$$G(s) = [Z(s)]^{-1}$$

$$= \frac{1}{\text{Det}} \begin{bmatrix} (2s^2 + 5s + 1) & 3s + 1 \\ 3s + 1 & (s^2 + 4s + 1) \end{bmatrix}$$

$$\text{Det} = (s^2 + 4s + 1)(2s^2 + 5s + 1) - (3s + 1)^2 = s(2s^3 + 13s^2 + 14s + 3)$$

EXAMPLE 2

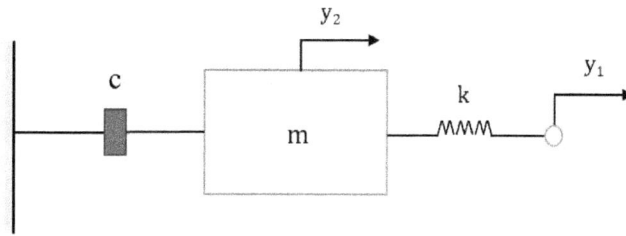

Introduce a virtual (dummy) mass m_0 at the degree of freedom, y_1.

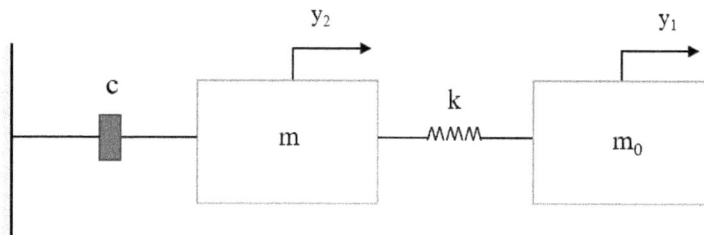

Let $m = 1\,\text{kg}$, $c = 1\text{kg/s}$, $k = 1\text{N/m}$, $m_0 = 0$:

$$[M] = \begin{bmatrix} 0 & 0 \\ 0 & 1 \end{bmatrix} \qquad [C] = \begin{bmatrix} 0 & 0 \\ 0 & 1 \end{bmatrix} \qquad [K] = \begin{bmatrix} 1 & -1 \\ -1 & 1 \end{bmatrix}$$

$$Z(s) = S^2[M] + S[C] + [K] = \begin{bmatrix} 1 & -1 \\ -1 & (s^2 + s + 1) \end{bmatrix}$$

Transfer Function

$$G(s) = [Z(s)]^{-1}$$

$$= \frac{1}{\text{Det}} \begin{bmatrix} (s^2 + s + 1) & 1 \\ 1 & 1 \end{bmatrix}$$

$$\text{Det} = (1)(s^2 + s + 1) - 1 = s(s + 1)$$

EXAMPLE 3

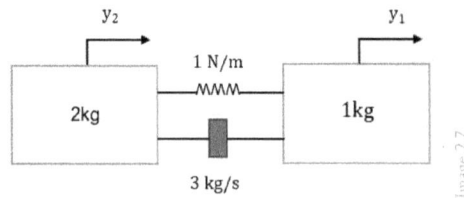

$$[M] = \begin{bmatrix} 1 & 0 \\ 0 & 2 \end{bmatrix} \qquad [C] = \begin{bmatrix} 3 & -3 \\ -3 & 3 \end{bmatrix} \qquad [K] = \begin{bmatrix} 1 & -1 \\ -1 & 1 \end{bmatrix}$$

$$Z(s) = S^2[M] + S[C] + [K] = \begin{bmatrix} (s^2 + 3s + 1) & -(3s+1) \\ -(3s+1) & (2s^2 + 3s + 1) \end{bmatrix}$$

Transfer Function

$$G(s) = [Z(s)]^{-1}$$

$$= \frac{1}{\text{Det}} \begin{bmatrix} (2s^2 + 3s + 1) & 3s+1 \\ 3s+1 & (s^2 + 3s + 1) \end{bmatrix}$$

$$\text{Det} = (2s^4 + 9s^3 + 3s^2 - 8)$$

2.5. Poles, Zeros, and the Final Value Theorem

Poles and zeros can be determined from a given TF. The steady-state value or final value of a system response can be easily obtained by applying the final value theorem.

2.5.1. Poles and Zeros

SINGLE DEGREE OF FREEDOM

$$m\ddot{y} + b\dot{y} + ky = 0$$

$$\ddot{y} + \frac{b}{m}\dot{y} + \frac{k}{m}y = 0$$

$$\ddot{y} + 2\varsigma\omega_n\dot{y} + \omega_n^2 y = 0$$

INITIAL CONDITIONS

$$y(0) = y_0 \qquad \dot{y}(0) = 0$$

$$\text{L.T.} => [s^2Y(s) - sy(0) - \dot{y}(0)] + (2\zeta\omega_n)[sY(s) - y(0)] + (\omega_n^2)[Y(s)] = 0$$

$$(s^2 + 2\zeta\omega_n s + \omega_n^2)Y(s) = (s + 2\zeta\omega_n)y_0$$

$$Y(s) = \frac{(s + 2\zeta\omega_n)y_0}{(s^2 + 2\zeta\omega_n s + \omega_n^2)} = \frac{N(s)}{D(s)}$$

Characteristic equation $\geq D(s) = 0$

The roots of this equation are called *poles*.

Numerator equation $\geq N(s) = 0$

The roots of this equation are called *zeros*.

With poles, $D(s) = 0 \geq Y(s) = \infty$

With zeros, $N(s) = 0 \geq Y(s) = 0$

The plot of poles and zeros is called the *S-plane plot*.

In this case, $Y(s)$ has one zero and two poles:

Zero: $S = -2\zeta\omega_n$

Poles: $S_{1,2} = -\zeta\omega_n \pm \omega_n\sqrt{\zeta^2 - 1}$

2.5.2. S-Plane Plots

$$S_{1,2} = -\zeta\omega_n \pm \omega_n\sqrt{\zeta^2 - 1}$$

Undamped: $\zeta = 0; \ S_{1,2} = \pm\omega_n\sqrt{-1} = \pm j\omega_n$

Critically damped: $\zeta = 1; \ S_{1,2} = -\omega_n$

Underdamped: $\zeta < 1; \ S_1$ and S_2 are complex

Overdamped: $\zeta > 1; \ S_1$ and S_2 are real

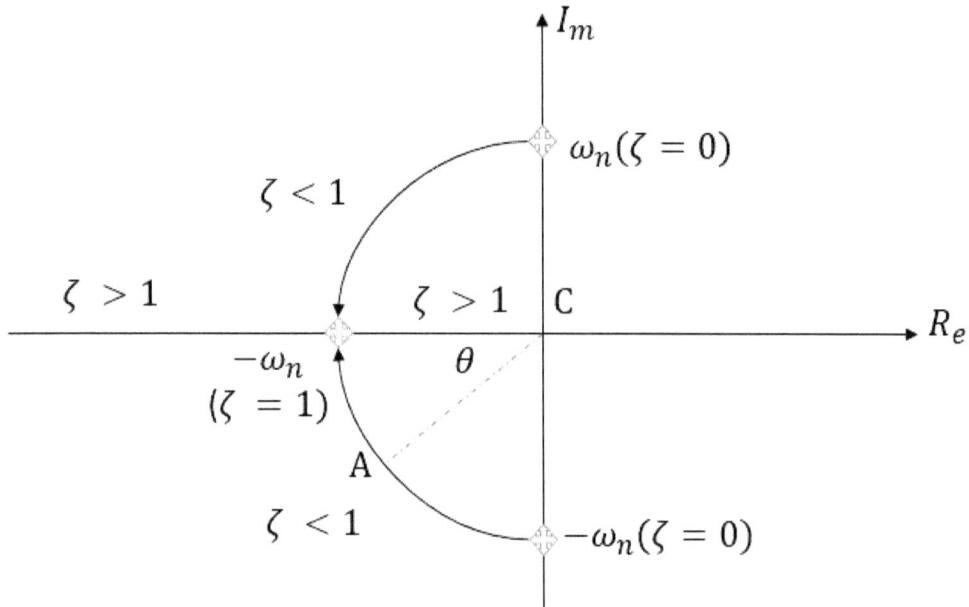

For $\zeta < 1$:

$$S_{1,2} = -\zeta\omega_n \pm j\omega_n\sqrt{1 - \zeta^2} = A \pm jB$$

$$\left|S\right| = \sqrt{A^2 + B^2} = \sqrt{\zeta^2\omega_n^2 + \omega_n^2\left(1 - \zeta^2\right)} = \omega_n = \overline{CA}$$

$$\tan\theta = \frac{B}{A} = \frac{\omega_n\sqrt{1 - \zeta^2}}{-\zeta\omega_n} = -\frac{\sqrt{1 - \zeta^2}}{\zeta}$$

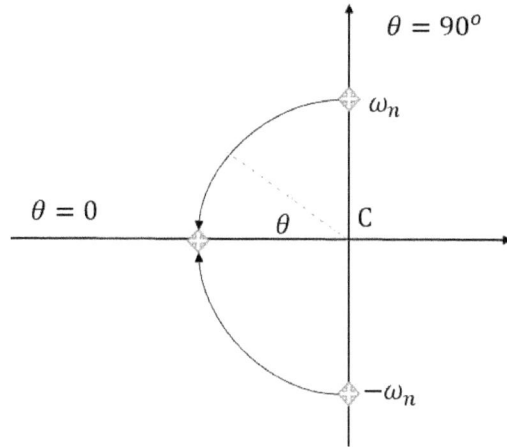

For:

ζ	0	1
θ	90	0

$$\sec^2\theta = 1 + \tan^2\theta = 1 + \frac{\left(1 - \zeta^2\right)}{\zeta^2} = \frac{1}{\zeta^2}$$

$$\sec\theta = \frac{1}{\zeta}; \ \cos\theta = \zeta; \theta = \cos^{-1}\zeta$$

2.5.3. Final Value Theorem

The final value (or steady state) of $y(t)$ is given by:

$$\lim_{t\to\infty} y(t) = \lim_{s\to 0} sY(s)$$

EXAMPLE

$$\ddot{y} + 4\dot{y} + 3y = 2r(t)$$

$$y(0) = 1, \ \dot{y}(0) = 0, \ r(t) = 1, \ t \geq 0 \ \text{(unit step function)}$$

LT

$$[s^2y(s) - sy(0) - \dot{y}(0)] + 4[sy(s) - y(0)] + 3y(s) = \frac{2}{s}$$

$$y(s)(s^2 + 4s + 3) = \frac{2}{s} + (s + 4)$$

$$y(s) = \frac{2}{s(s+3)(s+1)} + \frac{(s+4)}{(s+3)(s+1)}$$

$$= \frac{2 + 4s + s^2}{s(s+3)(s+1)}$$

Steady-state value:

$$\lim_{t \to \infty} y(t) = sY(s) \big|_{s=0}$$

$$= \frac{2 + 4s + s^2}{(s+3)(s+1)} \bigg|_{s=0} = \frac{2}{3}$$

EXAMPLE

$$\ddot{y} + 3\dot{y} + 2y = 1$$

$$y(0) = 1, \dot{y}(0) = 0$$

LT

$$[s^2y(s) - sy(0) - \dot{y}(0)] + 3[sy(s) - y(0)] + 2y(s) = \frac{1}{s}$$

$$y(s)[s^2 + 3s + 2] = \frac{1}{s} + (s + 3)$$

$$y(s) = \frac{1}{s(s^2 + 3s + 2)} + \frac{s+3}{s^2 + 3s + 2}$$

$$= \frac{s^2 + 3s + 1}{s(s^2 + 3s + 2)}$$

Final value:

$$\lim_{t \to \infty} y(t) = sY(s) \Big|_{s = 0}$$

$$= \frac{s^2 + 3s + 1}{s^2 + 3s + 2} \Big|_{s = 0} = \frac{1}{2}$$

2.5.4. Initial Value Theorem

$$\lim_{t \to 0}[y(t)] = \lim_{s \to \infty}\left[sY(s)\right] = y(0)$$

However, $y(0)$ is the initial condition.

In the above example:

$$\lim_{s \to \infty}\left[\frac{1 + 3s + s^2}{s^2 + 3s + 2}\right] = \left[\frac{\dfrac{1}{s^2} + \dfrac{3}{s} + 1}{1 + \dfrac{3}{s} + \dfrac{2}{s^2}}\right]_{s=\infty} = 1 = y(0)$$

2.6. Summary

The rule-based method helps for learning a quick and effective approach to develop the equations of motion for a system with multiple degrees of freedom. The equations of motion can be directly represented in a matrix format from a given system. However, this method is useful only for dynamic systems without inertial coupling. In other words, the mass matrix should be diagonal. The LT is applied to the equation of motion to obtain the TF. The poles are obtained by finding the roots of the denominator polynomial, whereas the zeros are obtained by finding the roots of the numerator polynomial of the TF. The poles and zeros can be represented in an S-plane plot to learn if the system is critically damped, overdamped, or underdamped. It is shown that the steady-state value of the response, y, at $t = \infty$ can be obtained by applying the final value theorem. Similarly, the initial value or initial condition at $t = 0$ can be obtained by applying the initial value theorem.

2.7. Assessment

1. The $[Z(s)]$ matrix is obtained from:

    a. $S^2[M] + S^2[C] + S^2[K]$

    b. $S[M] + S^2[C] + S^3[K]$

    c. $[M] + S[C] + S^2[K]$

    d. $S^2[M] + S[C] + [K]$

2. The TF G(s) can be obtained from:

    a. $[Z(s)]$

    b. $[Z(s)]^{-1}$

    c. $[Z(s)][Z(s)]$

    d. None of the above

3. The rule-based method is applicable when:

    a. The mass matrix is dynamically uncoupled

    b. The mass matrix is diagonal

    c. The stiffness and damping matrix are symmetric

    d. All of the above

4. Poles are defined as:

    a. The roots of the characteristic equation

    b. The square root of the characteristic equation

    c. The roots of the numerator equation of the TF

    d. The square root of the numerator equation

5. Zeros are defined as:

    a. The square root of the characteristic equation

    b. The square root of the numerator equation

    c. The roots of the numerator equation

    d. The roots of the denominator equation

6. With poles, the output $y(s)$ is equal to:

    a. Zero

    b. Infinity

    c. One

    d. None of the above

7. With zeros, the output $y(s)$ is equal to:

    a. Zero

    b. Infinity

    c. One

    d. None of the above

8. If $S_{1,2} = -\zeta w_n \pm j w_d$, the magnitude is equal to:

    a. The natural frequency

    b. The damped natural frequency

    c. The damping ratio

    d. None of the above

9. If $S_{1,2} = -\zeta w_n \pm j w_d$, the angle is equal to:

    a. $\cos^{-1}\zeta$

    b. $\cos\zeta$

    c. $\tan^{-1}\zeta$

    d. $\sin\zeta$

10. The final value theorem is useful to find:

    a. The value of output at $t = 0$

    b. The value of output at $t = \infty$

    c. The value of input at $t = 0$

    d. The value of input at $t = \infty$

2.8. Practice Problems

1. Answer all the assessment questions in Section 2.7

2. Determine the equation of motion for a dynamic system given below

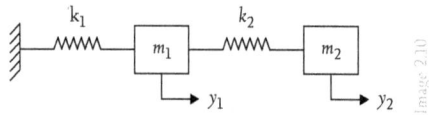

3. Determine the equation of motion for a dynamic system given below

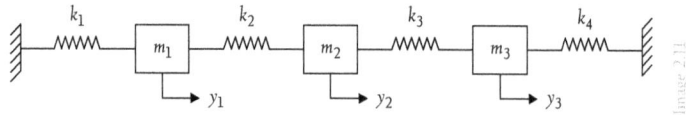

4. Determine the equation of motion for a dynamic system given below

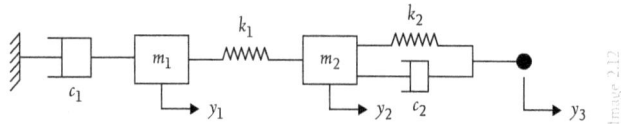

5. Determine the transfer function of a dynamic system given below, assuming $m_1 = 1$kg, $m_2 = 2$kg, $k = 1$N/m, $c_1 = 1$kg/s, $c_2 = 2$kg/s, $c_3 = 3$kg/s.

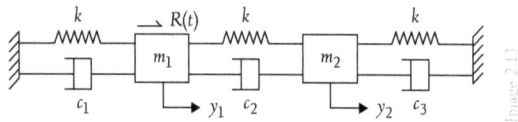

6. Determine the transfer function of a dynamic system given below, assuming $m_1 = 2$kg, $m_2 = 3$kg, $m_3 = 4$kg, $k_1 = 5$N/m, $k_2 = 10$N/m, $k_3 = 15$N/m, $k_4 = 20$N/m, $k_5 = 25$N/m, $c_1 = 5$kg/s, $c_2 = 6$kg/s (Hint: Use MATLAB)

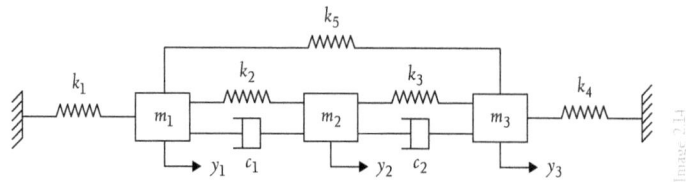

7. Determine the Poles and Zeros of the equation,

$$5\ddot{y} + 9\dot{y} + 45y = 0, \quad y(0) = 2, \qquad \dot{y}(0) = 1.4$$

8. Determine the Poles and Zeros of the equation and draw the S-Plane plot.

$$10\ddot{y} + 30\dot{y} + 250y = 0, \quad y(0) = 1, \qquad \dot{y}/(0) = 0$$

9. Determine the response $y(t)$ for given input of unit step function. Also find the steady state value using the final value theorem.

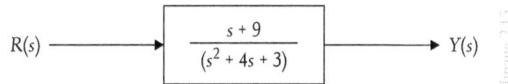

$$R(s) \longrightarrow \boxed{\dfrac{s+9}{(s^2 + 4s + 3)}} \longrightarrow Y(s)$$

10. Determine the response $y(t)$ for given input of (a) impulse function, (b) ramp function. Also plot the response for the range, $0 < t \leq 5$

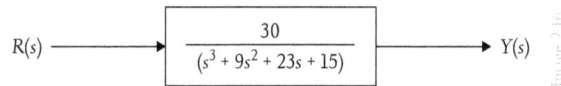

$$R(s) \longrightarrow \boxed{\dfrac{30}{(s^3 + 9s^2 + 23s + 15)}} \longrightarrow Y(s)$$

11. **MATLAB based problems:**

Using MATLAB, determine the transfer function for the following dynamic systems. Use [M], [C], [K] as input to MATLAB code.

a. $M_1 = 2\text{kg}$, $M_2 = 4\text{kg}$; $k_1 = 1\text{N/m}$, $k_2 = 3\text{N/m}$, $k_3 = 5\text{N/m}$; $b = 6\text{kg/s}$

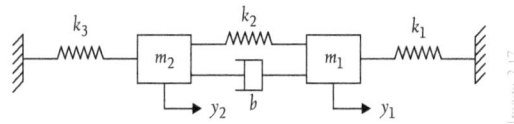

b. $M_1 = 1\text{kg}$, $M_2 = 2\text{kg}$, $M_3 = 3\text{kg}$; $k_1 = 1\text{N/m}$, $k_2 = 2\text{N/m}$, $k_3 = 3\text{N/m}$;
$k_4 = 4\text{N/m}$, $k_5 = 5\text{N/m}$; $b_1 = 1\text{kg/s}$, $b_2 = 2\text{kg/s}$, $b_3 = 3\text{kg/s}$, $b_4 = 4\text{kg/s}$, $b_5 = 5\text{kg/s}$

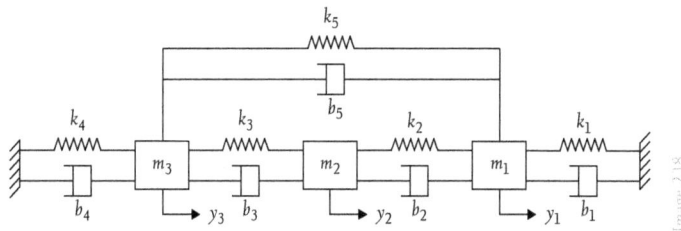

c. $M = 1\text{kg}; k = 1\text{N/m}$

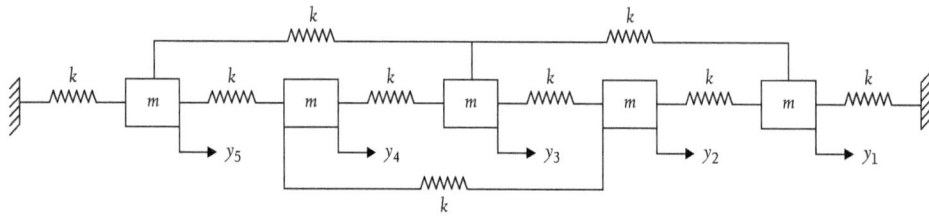

12. **SIMULINK based problem:**

$$a\ddot{x} + b\dot{x} + cx = p\dot{y} + qy$$

Using SIMULINK, determine the response $x(t)$ for given input $y(t)$.

The constants, $a = 1, b = 2, c = 3, p = 4, q = 5$

The input function $y(t)$ is given as,

a. Step function

b. Ramp function

c. Impulse function

d. Sinusoidal

3

MODELING OF CONTROL SYSTEMS

3.1. Introduction

The control system can be modeled as a group of blocks connected together as a block diagram or group of nodes connected by branches as a signal flow graph. In a block diagram, several rules are included to reduce the number of blocks into a single equivalent block to find the total transfer function (TF). In a signal flow graph, the total TF of the system is obtained by using Mason's formula. The process for applying both of these methods is presented in this chapter.

3.2. Learning Objectives

1. Model the control system with a block diagram and determine the TF for the total system.

2. Reduce the given block diagram with multiple blocks into a single equivalent block.

3. Model the control system with a signal flow graph and determine the TF for the entire system.

4. Find the path TFs and loop TFs to apply Mason's formula.

3.3. Block Diagram Model

A control system would have several blocks, and each block has a TF. However, the TF for the entire system can be achieved by minimizing all the blocks into a single equivalent block. The resulting TF would give the required TF for the whole system. Minimizing the blocks is based on a set of rules, which are provided below.

3.3.1. Rules for Minimizing the Number of Blocks

1. **Blocks in Series**

R $\longrightarrow$ G_1 $\longrightarrow$ G_2 $\longrightarrow$ Y $\implies$ R $\longrightarrow$ G_1G_2 $\longrightarrow$ Y

Image 3.1a&b

2. **Feedback Loop**

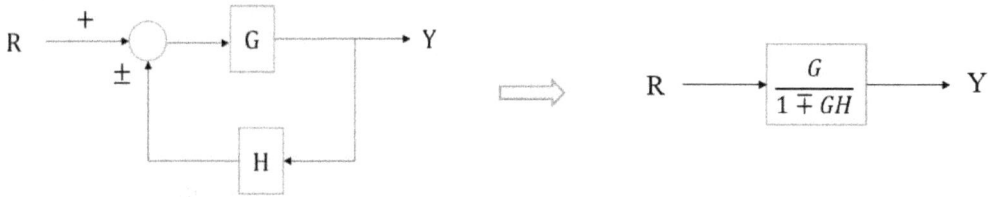

3. **Moving Summing Point**

   a. Ahead to behind

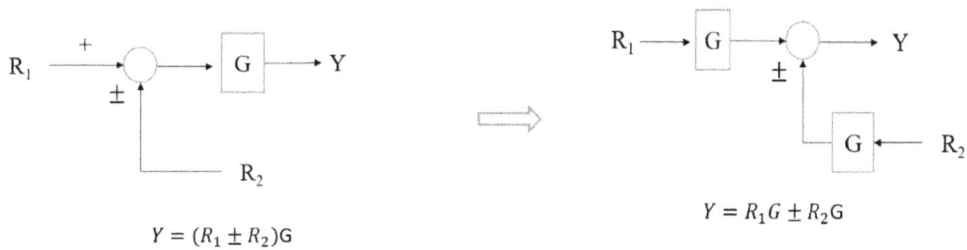

$$Y = (R_1 \pm R_2)G$$

$$Y = R_1 G \pm R_2 G$$

   b. Behind to ahead

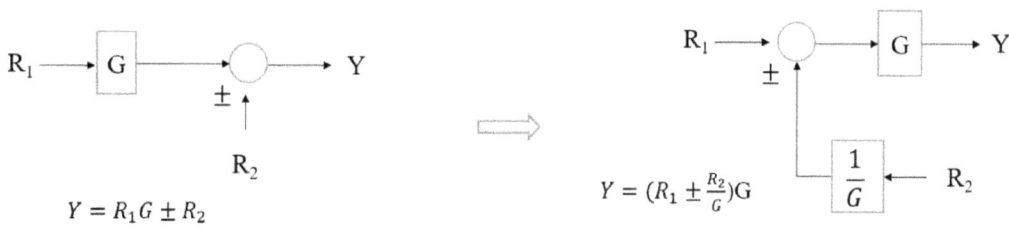

$$Y = R_1 G \pm R_2$$

$$Y = (R_1 \pm \frac{R_2}{G})G$$

4. **Moving Pick-Off Point**

   a. Ahead to behind

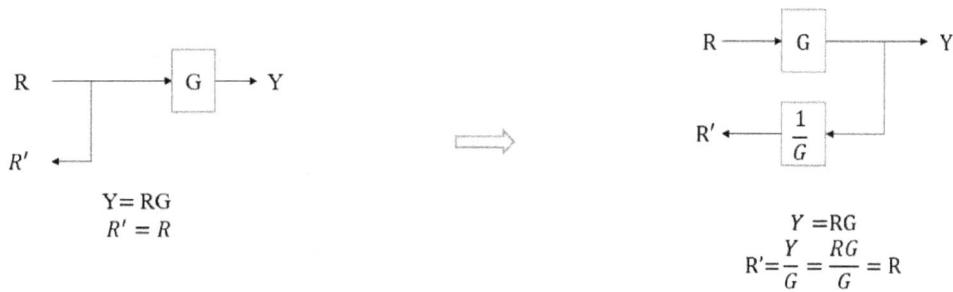

$$Y = RG$$
$$R' = R$$

$$Y = RG$$
$$R' = \frac{Y}{G} = \frac{RG}{G} = R$$

b. Behind to ahead

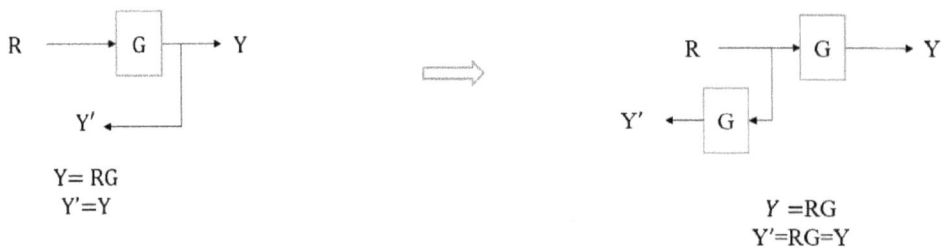

Y= RG
Y'=Y

$\Longrightarrow$

$Y = RG$
Y'=RG=Y

3.3.2. Equivalent Block Diagram

EXAMPLE
Minimize the given block diagram to a single equivalent block:

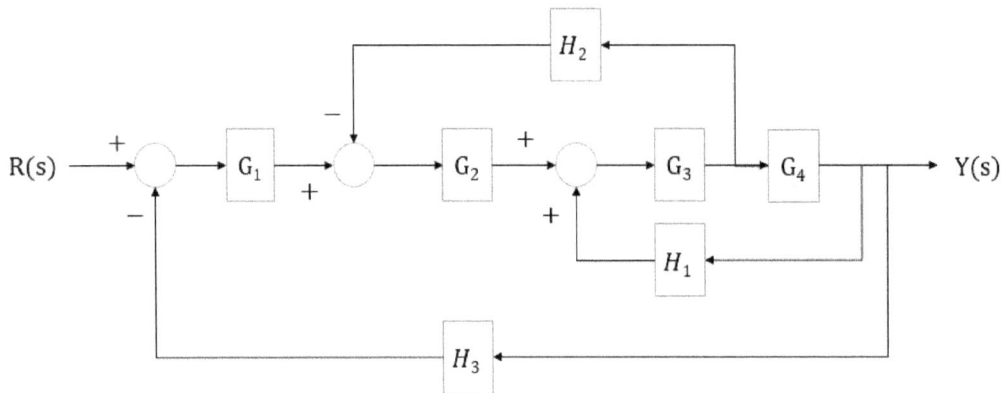

Step 1: Move the pick-off point from ahead of G_4 to behind G_4:

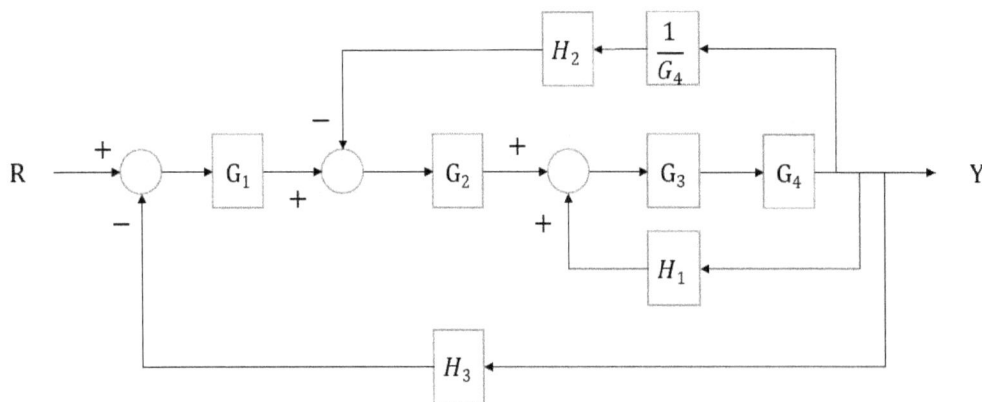

Step 2: Minimize the feedback loop of $G_3G_4H_1$:

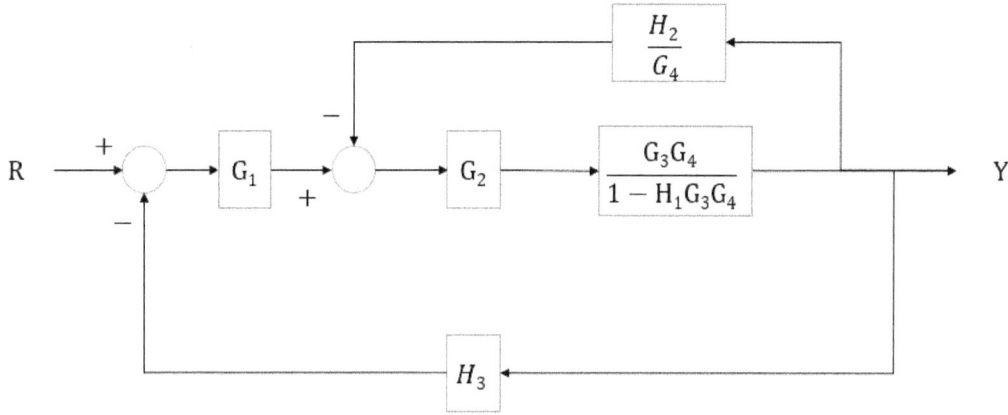

Step 3:

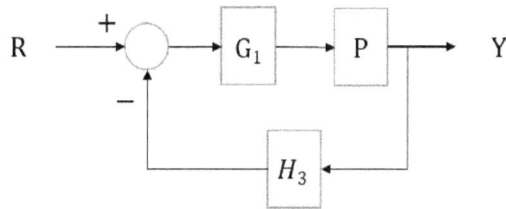

Let $K_1 = \left(\dfrac{G_3G_4}{1 - G_3G_4H_1} \right)$, $K_2 = \left(\dfrac{H_2}{G_4} \right)$ and minimize the feedback loop of $G_2K_1K_2$:

$$ P = \frac{G_2K_1}{1 + G_2K_1K_2} $$

Step 4: Minimize the feedback G_1, P, H_3:

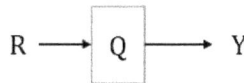

$$ Q = \frac{G_1P}{1 + G_1H_3P} $$

$$P = \frac{G_2 G_3 G_4}{1 - G_3 G_4 H_1} \bigg/ 1 + G_2 \frac{H_2}{G_4} \frac{G_3 G_4}{(1 - G_3 G_4 H_1)}$$

$$= \frac{G_2 G_3 G_4}{1 - G_3 G_4 H_1} \bigg/ \left\{ \frac{1 - G_3 G_4 H_1 + G_2 G_3 H_2}{(1 - G_3 G_4 H_1)} \right\}$$

$$= \frac{G_2 G_3 G_4}{(1 - G_3 G_4 H_1)} \frac{(1 - G_3 G_4 H_1)}{(1 - G_3 G_4 H_1 + G_2 G_3 H_2)}$$

$$= \frac{G_2 G_3 G_4}{1 - G_3 G_4 H_1 + G_2 G_3 H_2}; \; P = \frac{G_2 G_3 G_4}{D}$$

$$D = 1 - G_3 G_4 H_1 + G_2 G_3 H_2$$

$$G_1 P = \frac{G_1 G_2 G_3 G_4}{D}$$

$$1 + G_1 P H_3 = 1 + \frac{G_1 G_2 G_3 G_4 H_3}{D}$$

$$1 + G_1 P H_3 = \frac{D + G_1 G_2 G_3 G_4 H_3}{D}$$

$$Q = \frac{G_1 P}{1 + G_1 P H_3} = \frac{G_1 G_2 G_3 G_4}{D + G_1 G_2 G_3 G_4 H_3} = \frac{G_1 G_2 G_3 G_4}{(1 - G_3 G_4 H_1 + G_2 G_3 H_2 + G_1 G_2 G_3 G_4 H_3)}$$

EXAMPLE

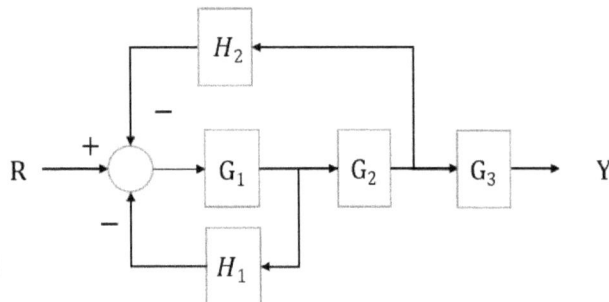

Step 1: Minimize the feedback loop G_1, H_1.

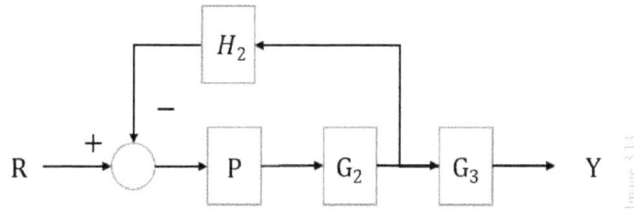

$$P = \frac{G_1}{1 + G_1 H_1}$$

Step 2: Minimize the feedback loop P, G_2, H_2.

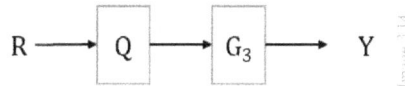

$$Q = \frac{PG_2}{1 + PG_2 H_2} = \frac{G_1 G_2}{1 + G_1 H_1 + G_1 G_2 H_2}$$

Step 3: Minimize the Q, G_3 blocks in the series:

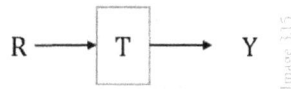

$$T = \frac{Y}{R} = G_3 Q = \frac{G_1 G_2 G_3}{(1 + G_1 H_1 + G_1 G_2 H_2)}$$

3.3.3. Alternate Solution

This method is alternate to applying to the rules to minimize the blocks. In this method, the output for each block is determined and then consolidated to find the TF for a single equivalent block.

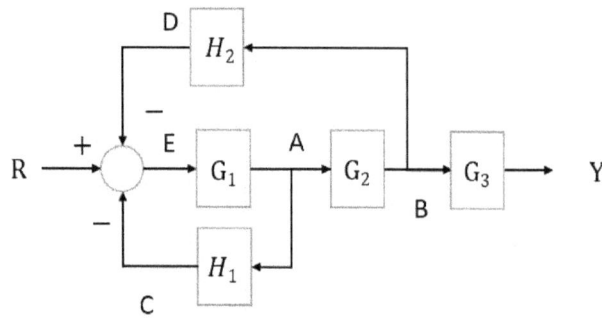

Step 1: Label the output from each block and summing points.

Step 2: Write the equation output = (input) (TF) for each block and simplify to find $Y = (T)(R)$, where $T(s)$ is the total or equivalent TF.

$$E = R - C - D; \ C = AH_1; \ D = BH_2; \ E = R - AH_1 - BH_2$$

$$A = EG_1 = (R - AH_1 - BH_2)G_1; \ A(1 + G_1H_1) = (R - BH_2)G_1$$

$$A = \frac{(R - BH_2)G_1}{1 + G_1H_1}$$

$$B = AG_2 = \frac{(R - BH_2)G_1G_2}{1 + G_1H_1}; \ B(1 + G_1H_1) = RG_1G_2 - BH_2G_1G_2$$

$$B(1 + G_1H_1 + G_1G_2H_2) = RG_1G_2; \ B = \frac{RG_1G_2}{1 + G_1H_1 + G_1G_2H_2}$$

$$Y = BG_3 = \frac{RG_1G_2G_3}{1 + G_1H_1 + G_1G_2H_2}$$

$$T = \frac{Y}{R} = \frac{G_1G_2G_3}{1 + G_1H_1 + G_1G_2H_2}$$

3.4. Signal Flow Graph Models

The signal flow graph is also a viable method for modeling the control system. In this method, the input and output for each block is represented by nodes designated by a circle, and the TF is represented by branches designated by a line. The graph consists of nodes connected by branches.

Block Diagram   **Signal Flow Graph**

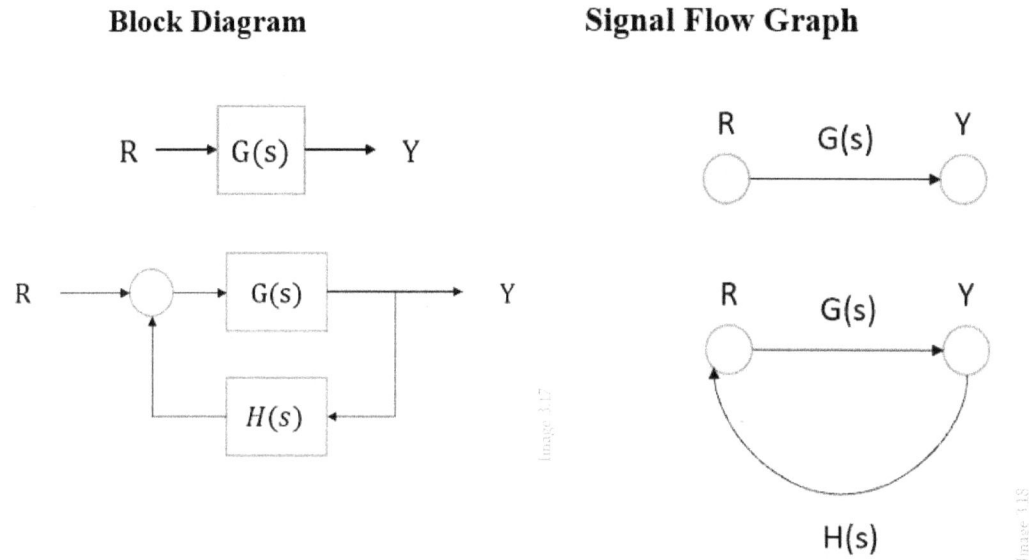

In a signal flow graph, the description of parameters is given below.

Nodes: input and output points or junctions

Path (or branch): line connecting one signal node with another signal node; represents the TF (or the ratio of output to input)

Loop: closed path that starts from output node and end at input node; represents feedback in the control system

Signal flow gain: $T(s) = \dfrac{Y(s)}{R(s)}$

The relation between paths and signal flow gain (or TF) is given by Mason's formula:

$$T = \frac{\Sigma(P_k \Delta_k)}{\Delta}$$

Where P_k = Product of gains on k^{th} path from input, R(s) to output, Y(s)

$$\Delta = \text{determinant} = 1 - \sum_{n=1}^{N} L_n + \Sigma L_n L_m - \Sigma L_n L_m L_p + \cdots\cdots$$

L_n = loop gain, N = number of loops

$L_n L_m$ = product of gains of all combinations of two loops that are not touching each other

$L_n L_m L_p$ = product of gains of all combinations of three loops that are not touching each other
Δ_k = cofactor of the path P_k

= the value of Δ, after removing the loops, that are touching the kth path (or the value of Δ reduced by setting the loops touching the kth path to be zero)

3.4.1. Single Path
There is only one path connecting input (R) and output (Y):

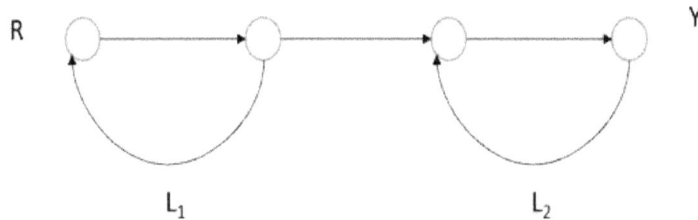

$$N = 2 \text{ loops}; \ L_1 \text{ and } L_2$$

$$\Delta = 1 - \sum_{n=1}^{2} L_n + \sum_{n=1}^{2} L_n L_m$$

$$= 1 - (L_1 + L_2) + L_1 L_2$$

L_1 and L_2 are non-touching loops; they do not touch each other. So $L_1 L_2 \neq 0$.
Because there is only one path, $k = 1$. So $\Delta_k = \Delta_1 = 1$, because the loops L_1 and L_2 touch the path. $L_1 = 0 = L_2$.

OVERLAPPING LOOP

Because L_1 and L_2 overlap, these loops are assumed to be touching each other. So $L_1 L_2 = 0$.

$$\Delta = 1 - (L_1 + L_2) \qquad \Delta_1 = 1$$

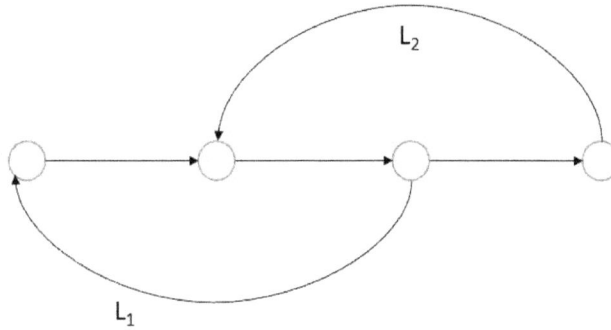

SAMPLE FOR TOUCHING LOOPS

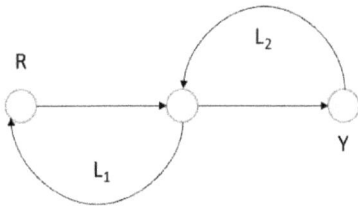

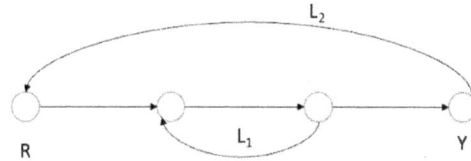

$\Delta = 1 - (L1 + L2) \qquad \Delta_1 = 1$
$\Delta = 1 - (L1 + L2) \qquad \Delta_1 = 1$

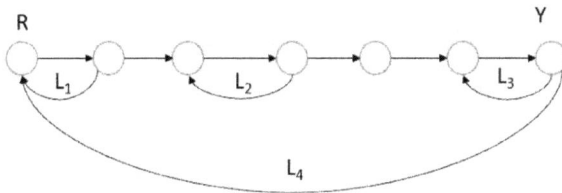

L_1, L_2, L_3 → do not touch each other. L_4 overlaps L_1, L_2, L_3; hence, it touches all other loops.

$$\Delta = 1 - (L_1 + L_2 + L_3 + L_4) + (L_1 L_2 + L_2 L_3 + L_1 L_3) - L_1 L_2 L_3$$

$$\Delta_1 = 1 \text{ (Because } L_1, L_2, L_3, \text{ and } L_4 \text{ touch the path)}$$

3.4.2. Multiple Paths

TWO PATHS

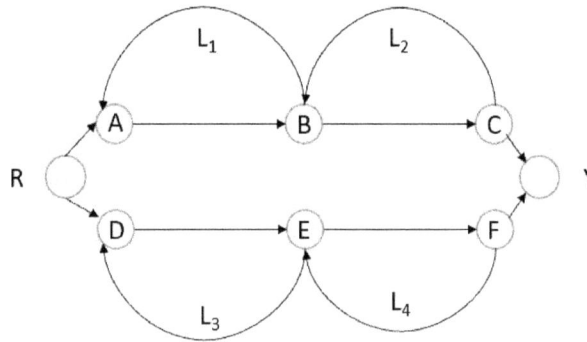

Path 1 = R-A-B-C-Y

Path 2 = R-D-E-F-Y

$k = 1, 2$ (2 paths)

$N = 4$ loops

L_1 and L_2 are touching, and L_3 and L_4 are touching; $L_1 L_2 = 0 = L_3 L_4$.

L_1 and L_2 do not touch L_3 and L_4.

$$\text{So, } \sum L_n L_m = L_1 L_3 + L_1 L_4 + L_2 L_3 + L_2 L_4$$

$$\sum L_n = L_1 + L_2 + L_3 + L_4; \sum L_n L_m L_P = 0$$

$$\Delta = 1 - \sum L_n + \sum L_n L_m$$

$$= 1 - (L_1 + L_2 + L_3 + L_4) + (L_1 L_3 + L_1 L_4 + L_2 L_3 + L_2 L_4)$$

$$k = 1 \qquad \Delta_1 = \Delta \big|_{L_1 = 0, L_2 = 0} = 1 - (L_3 + L_4) + (0)$$

$$k = 2 \qquad \Delta_2 = \Delta \big|_{L_3 = 0, L_4 = 0} = 1 - (L_1 + L_2) + (0)$$

THREE PATHS

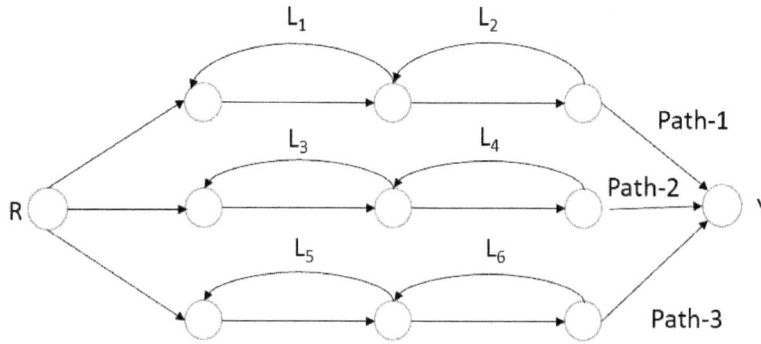

$N = 6$

$k = 1, 2, 3$

Path 1: $L_1 = 0 = L_2$

Path 2: $L_3 = 0 = L_4$

Path 3: $L_5 = 0 = L_6$

TOUCHING LOOPS: $L_1 L_2 = 0, L_3 L_4 = 0, L_5 L_6 = 0$

$$\Delta = 1 - \sum L_n + \sum L_n L_m - \sum L_n L_m L_p$$

$$\sum L_n = L_1 + L_2 + L_3 + L_4 + L_5 + L_6$$

$$\sum L_n L_m = L_1 L_3 + L_1 L_4 + L_1 L_5 + L_1 L_6 + L_2 L_3 + L_2 L_4 + L_2 L_5 + L_2 L_6 + L_3 L_5 + L_3 L_6 + L_4 L_5 + L_4 L_6$$

$$\sum L_n L_m L_p = L_1 L_3 L_5 + L_1 L_3 L_6 + L_2 L_4 L_5 + L_2 L_4 L_6 + L_1 L_4 L_5 + L_1 L_4 L_6 + L_2 L_3 L_5 + L_2 L_3 L_6$$

$$k = 1: \Delta_1 = \Delta \big|_{L_1 = 0, L_2 = 0} = 1 - (L_3 + L_4 + L_5 + L_6) + (L_3 L_5 + L_3 L_6 + L_4 L_5 + L_4 L_6)$$

$$k = 2: \Delta_2 = \Delta \big|_{L_3 = 0, L_4 = 0} = 1 - (L_1 + L_2 + L_5 + L_6) + (L_1 L_5 + L_1 L_6 + L_2 L_5 + L_2 L_6)$$

$$k = 3: \Delta_3 = \Delta \big|_{L_5 = 0, L_6 = 0} = 1 - (L_1 + L_2 + L_3 + L_4) + (L_1 L_3 + L_1 L_4 + L_2 L_3 + L_2 L_4)$$

3.4.3. Equivalent Signal Flow Graph Model

For a given block diagram model, we can find the signal flow graph model and determine the total TF. The block diagram of the first example in Section 3.3.2 is given below.

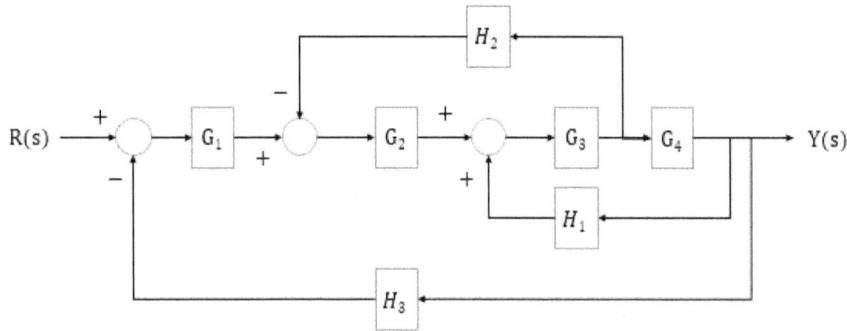

The above block diagram model can be shown as an equivalent signal flow graph model as given below.

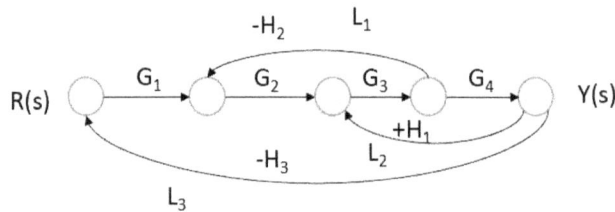

Number of loops: $N = 3$ (all are touching each other)

Number of paths: $k = 1$ (only one path from $R(s)$ to $Y(s)$)

$$T(s) = \frac{Y(s)}{R(s)} = \frac{P_1 \Delta_1}{\Delta} \qquad \text{(Mason's formula)}$$

Path gain: $P_1 = G_1 G_2 G_3 G_4$ = product of TF along the path from input to output.

$$\Delta = 1 - \sum L_n + \sum L_n L_m$$

$$\sum L_n = L_1 + L_2 + L_3$$

$$\sum L_n L_m = L_1 L_2 + L_1 L_3 + L_2 L_3 = 0$$

The loops L_1, L_2, and L_3 overlap, and hence, they touch each other. So, the product $(L_n L_m)$ is zero.

$$\Delta = 1 - (L_1 + L_2 + L_3); \Delta_1 = 1$$

For a single path, all the loops touch the path and hence set $L_1 = 0$, $L_2 = 0$, $L_3 = 0$ in Δ.

Loop TF = product of TF along the loop.

$L_1 = -G_2 G_3 H_2$

$L_2 = G_3 G_4 H_1$

$L_3 = -G_1 G_2 G_3 G_4 H_3$

$$T(s) = \frac{G_1 G_2 G_3 G_4}{(1 + G_2 G_3 H_2 - G_3 G_4 H_1 + G_1 G_2 G_3 G_4 H_3)}$$

EXAMPLE

For the given signal flow graph, determine the TF:

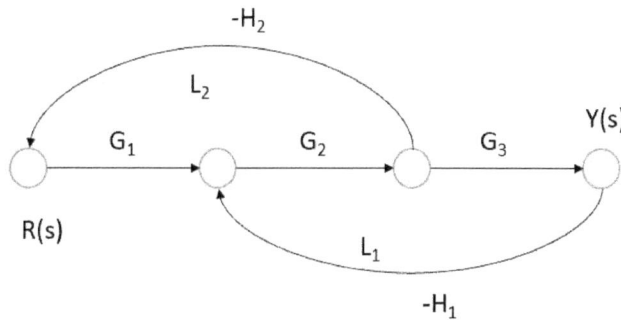

$$T(s) = \frac{P_1 \Delta_1}{\Delta}$$

Number of paths = 1; $k = 1$

Path gain: $P_1 = G_1 G_2 G_3$

Number of loops = 2; L_1 and L_2

Loop gain: $L_1 = -G_2 G_3 H_1$ and $L_2 = -G_1 G_2 H_2$

$$\Delta = 1 - \sum L_n + \sum L_n L_m$$

$$\sum L_n = L_1 + L_2$$

$$\sum L_n L_m = L_1 L_2 = 0$$

Because the loops are touching.

$$\Delta = 1 - (L_1 + L_2)$$

$$\Delta_1 = \Delta \big|_{L_1 = 0, L_2 = 0} = 1 \text{ (because both loops touch the path)}$$

$$T(s) = \frac{G_1 G_2 G_3}{(1 + G_2 G_3 H_1 + G_1 G_2 H_2)}$$

This signal flow graph model can be represented as a block diagram model as given below.

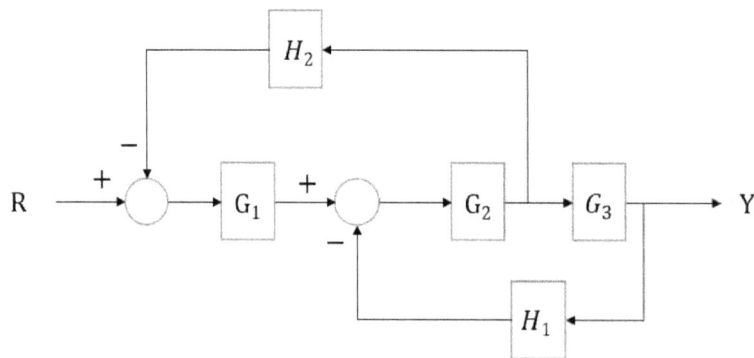

To minimize the above block diagram into an equivalent single block:

Step 1: Move the pick-off point from ahead of G_3 to behind G_3:

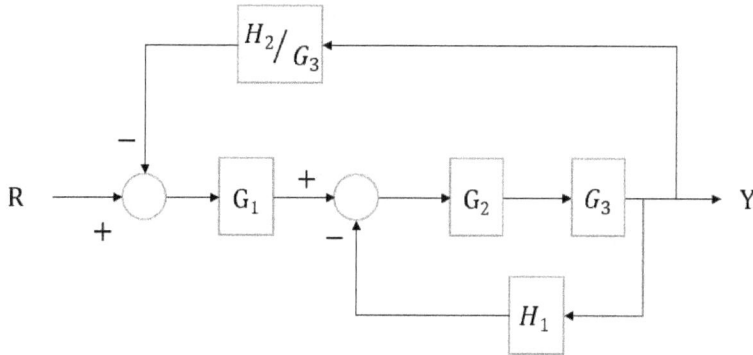

Step 2: Minimize the G_2, G_3, H_1 feedback loop:

$$P = \frac{G_2 G_3}{1 + G_2 G_3 H_1}$$

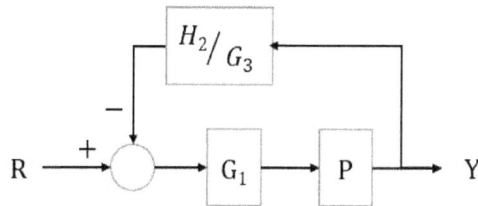

Step 3 Minimize the G_1, P, H_2 / G_3 feedback loop:

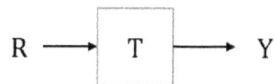

$$T = \frac{Y}{R} = \frac{G_1 P}{1 + (G_1 P H_2 / G_3)} = \frac{G_1 G_3 P}{G_3 + G_1 H_2 P}$$

$$= \frac{G_1 G_3 \left\{ \dfrac{G_2 G_3}{1 + G_2 G_3 H_1} \right\}}{G_3 \left\{ 1 + \dfrac{G_1 G_2 H_2}{1 + G_2 G_3 H_1} \right\}}$$

$$T = \frac{G_1 G_2 G_3}{(1 + G_2 G_3 H_1 + G_1 G_2 H_2)}$$

EXAMPLE

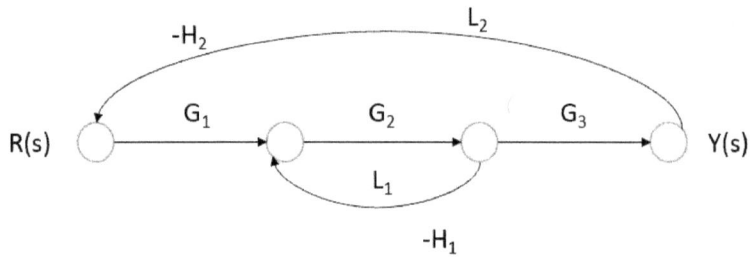

$$T(s) = \frac{p_1 \Delta_1}{\Delta}$$

$$P_1 = G_1 G_2 G_3$$

$$L_1 = -G_2 H_1 \text{ and } L_2 = -G_1 G_2 G_3 H_2$$

$$\Delta = 1 - \sum L_n = 1 - (L_1 + L_2)$$

$$\Delta_1 = 1$$

$$T(s) = \frac{G_1 G_2 G_3}{(1 + G_2 H_1 + G_1 G_2 G_3 H_2)}$$

EXAMPLE

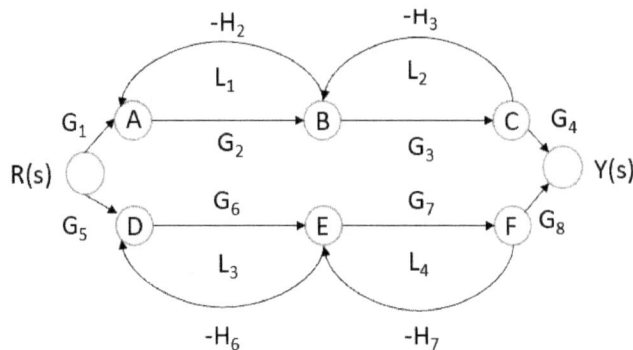

$$T(s) = \frac{\sum P_k \Delta_k}{\Delta} = \frac{1}{\Delta}(P_1 \Delta_1 + P_2 \Delta_2)$$

Number of paths = 2; $k = 1, 2$

Path gain:

$$P_1 = G_1 G_2 G_3 G_4 \quad \text{(upper path)}$$

$$P_2 = G_5 G_6 G_7 G_8 \quad \text{(lower path)}$$

Number of loops = 4

Loop gain: $L_1 = -G_2 H_2 ; L_2 = -G_3 H_3; L_3 = -G_6 H_6 ; L_4 = -G_7 H_7$

Nontouching loops: Loops L_1 and L_2 do not touch L_3 and L_4:

$$\Delta = 1 - \sum L_n + \sum L_n L_m$$

$$\sum L_n = L_1 + L_2 + L_3 + L_4 = -(G_2 H_2 + G_3 H_3 + G_6 H_6 + G_7 H_7)$$

$$\sum L_n L_m = L_1 L_3 + L_1 L_4 + L_2 L_3 + L_2 L_4$$

$$(L_1 L_2 = 0 = L_3 L_4, \text{ because they are touching the loops})$$

$$\Delta = 1 - (L_1 + L_2 + L_3 + L_4) + (L_1 L_3 + L_1 L_4 + L_2 L_3 + L_2 L_4)$$

$$\Delta_1 = \Delta \big|_{L_1 = 0 = L_2} = 1 - (L_3 + L_4) = 1 + G_6 H_6 + G_7 H_7$$

$$\Delta_2 = \Delta \big|_{L_3 = 0 = L_4} = 1 - (L_1 + L_2) = 1 + G_2 H_2 + G_3 H_3$$

$$T(s) = \frac{(G_1 G_2 G_3 G_4)(1 + G_6 H_6 + G_7 H_7) + (G_5 G_6 G_7 G_8)(1 + G_2 H_2 + G_3 H_3)}{[1 + (G_2 H_2 + G_3 H_3 + G_6 H_6 + G_7 H_7) + (G_2 G_6 H_2 H_6 + G_2 G_7 H_2 H_7 + G_3 G_6 H_3 H_6 + G_3 G_7 H_3 H_7)]}$$

3.5. Summary

The rules for minimizing the number of blocks in a series, an equivalent block for a feedback loop, and an equivalent block for moving a summing or pick-off point are shown in this chapter, with examples provided. It helps to reduce the given system with several blocks into an equivalent single input, a single-output system with a single block. However, the total system's TF becomes complex. The signal flow graph method serves as an alternate method, where the input and output nodes are connected by branches that represent the TF. Here, Mason's formula is applied in finding the total TF. However, the application of this formula has some constraints in identifying the path gain, loop gain, determinant, and cofactor. Both methods are equally effective in modeling the control system.

3.6. Assessment

1. If G_1 and G_2 are in series, the equivalent TF is equal to:

   a. G_1/G_2

   b. $G_1 + G_2$

   c. $G_1 - G_2$

   d. $G_1{}^*G_2$

2. When you move the summing point from behind to ahead, the input R_2 becomes:

   a. $R_2 + G$

   b. R_2/G

   c. $R_2 - G$

   d. $R_2{}^*G$

3. When you move the pick-off point from ahead to behind, the pick-off value is:

   a. Divided by G

   b. Multiplied by G

   c. All of the above

   d. None of the above

4. In a signal flow graph, the nodes represent:

    a. Input

    b. Output

    c. Junction point in a path

    d. All of the above

5. In a signal flow graph, the lines (branches) represent:

    a. The total TF

    b. The TF of a block

    c. The loop TF

    d. The path TF

6. The path TF is equal to:

    a. The sum of all TFs in a path

    b. The ratio of all TFs in a path

    c. The product of all TFs in a path

    d. None of the above

7. Mason's formula is a function of:

    a. The path TF

    b. The loop TF

    c. The determinant function

    d. All of the above

8. For single paths:

    a. $\Delta_1 = 1$

    b. All of the loop touches the path

    c. $T(s) = P_1 \Delta_1 / \Delta$

    d. All of the above

9. For two paths:

   a. $T(s) = (P_1 + P_2)(\Delta_1 + \Delta_2)/\Delta$

   b. $T(s) = (P_1\Delta_1 + P_2\Delta_2)/\Delta$

   c. $T(s) = (P_1 - P_2)(\Delta_1 - \Delta_2)/\Delta$

   d. $T(s) = (P_1\Delta_1 - P_2\Delta_2)/\Delta$

10. The loops are identified as touching if:

    a. They are adjacent

    b. They cross

    c. They overlap

    d. All of the above

3.7. Practice Problems

1. Answer all the assessment questions in Section 3.6

2. Determine the system transfer function by minimizing the given block diagram into a single equivalent block. $G_1 = (s + 2)$; $G_2 = (1/s^2)$; $G_3 = (1/s)$; $G_4 = s(s - 1)$

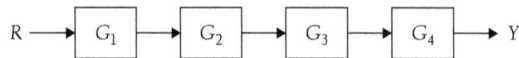

3. Move the summing block ahead of G_1 and draw the resulting block diagram.

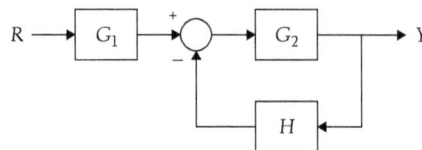

4. Move the pick-off point ahead of G_2 and draw the resulting block diagram. Determine the system transfer function before and after moving the pick-off point.

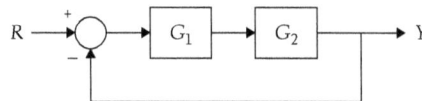

5. Determine the system transfer function by minimizing the given block diagram into a single equivalent block.

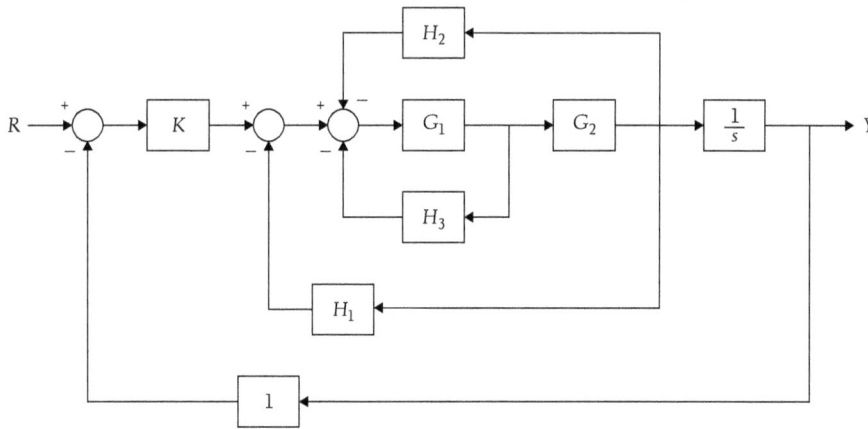

6. Determine the system transfer function by minimizing the given block diagram into a single equivalent block.

$$K = 4, G_1 = 1/(s + 1), G_2 = s/(s^2 + 2), G_3 = (1/s^2), H_1 = (4s + 2)/(s^2 + 2s + 1), H_2 = 50,$$
$$H_3 = (s^2 + 2)/(s^3 + 14)$$

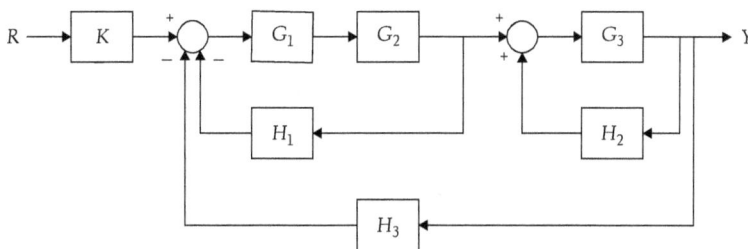

7. Determine the transfer function for the given signal flow graph model by Mason's formula.

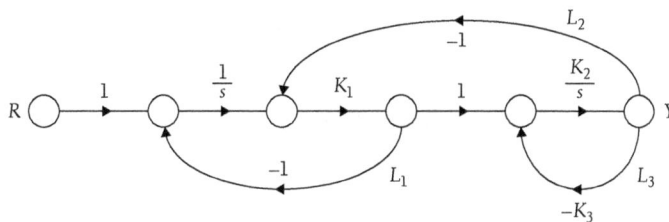

8. Determine the transfer function for the given signal flow graph model by Mason's formula.

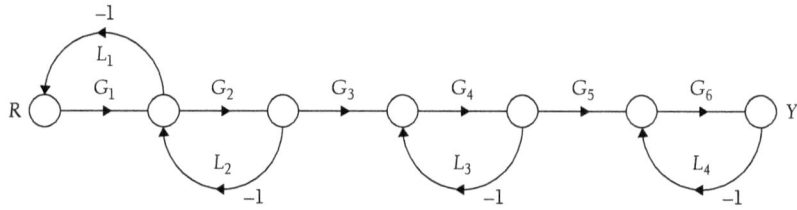

9. Draw the block diagram model for the signal flow graph model given below.

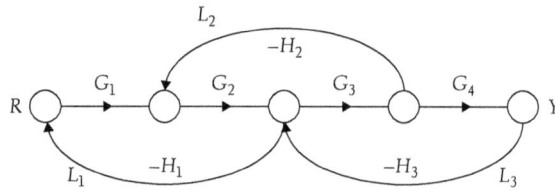

10. Determine the transfer function by Mason's formula for the multiple path signal flow graph model given below.

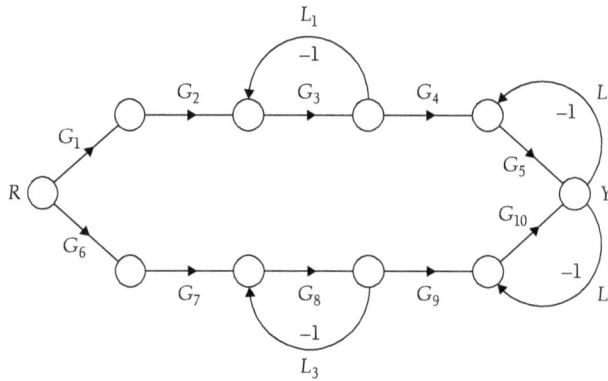

11. **SIMULINK based problems:**

Using SIMULINK determine the output function. Print the SIMULINK model and the output result.

a. For the control system given below, $R(t) = $ Unit Step, $K = 10$,

$G_1 = (s + 1)/(s + 5)$; $G_2 = 1/(s^2)$; $H = 1$

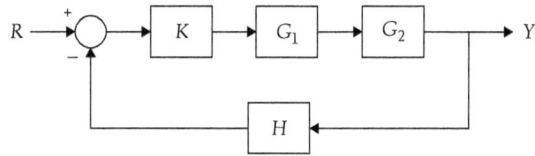

b. Determine the output function for three different input functions given as (i) Unit Impulse, (ii) Unit Step, and (iii) Ramp.

The Transfer function, $G(s) = (30)/(s^3 + 9 s^2 + 23s + 15)$

c. Determine the output function for three different input functions given as (i) Unit Impulse, (ii) Unit Step, and (iii) Ramp.

The Transfer function, $G(s) = (30)/(s^3 + 9 s^2 + 26s + 30)$

CHARACTERISTICS OF CONTROL SYSTEMS

4

4.1. Introduction

The feedback system generally has some differences between a given input and its measured output. This difference is called an *error in the system*. In addition, there may be an external disturbance or noise in measurement that may contribute to system error. In this chapter, emphasis is placed on studying errors and learning how to minimize them. The system sensitivity analysis is also presented to examine the effect of a specific parameter on system error. Also discussed is a method to obtain the steady-state error using the final value theorem. Applications for typical biomedical and aerospace systems are included.

4.2. Learning Objectives

1. Study the role of error signals in s negative-feedback control system.

2. Minimize the system's sensitivity to parameter changes and unwanted disturbance or noise.

3. Apply the final value theorem to determine the steady-state error.

4.3. Error Signal Analysis

An error is defined as the difference between the input (desired output) and the output (actual output). $E(s) = R(s) - Y(s)$

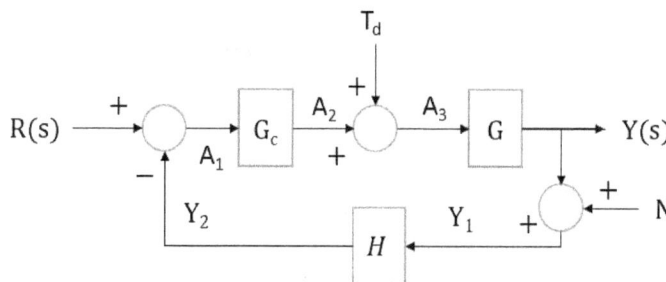

$N = $ noise

$T_d = $ disturbance

$H = 1$ (unit feedback)

$G = $ plant transfer function (TF)

G_c = controller TF

$Y_1 = Y + N$

$A_1 = R - Y_2;\ Y_2 = Y_1 H = Y_1$

$A_1 = R - Y_1 = R - Y - N$

$A_2 = A_1 G_c = G_c(R - Y - N)$

$A_3 = A_2 + T_d = T_d + G_c R - G_c Y - G_c N$

$Y = A_3 G = GT_d + GG_c R - GG_c Y - GG_c N$

$Y(1 + GG_c) = GT_d + GG_c R - GG_c N$

$$Y = \left(\frac{GG_c}{1 + GG_c}\right)R + \left(\frac{G}{1 + GG_c}\right)T_d - \left(\frac{GG_c}{1 + GG_c}\right)N$$

Let $L = GG_c$, the loop TF (forward loop):

$$Y = \left(\frac{L}{1 + L}\right)R + \left(\frac{G}{1 + L}\right)T_d - \left(\frac{L}{1 + L}\right)N$$

Tracking error: $E(s) = R(s) - Y(s) = R\left(\frac{1}{1 + L}\right) - \left(\frac{G}{1 + L}\right)T_d + \left(\frac{L}{1 + L}\right)N$

Let $S = \dfrac{1}{1 + L}$ = sensitivity function

$C = \dfrac{L}{1 + L}$ = complementary sensitivity function

$$= 1 - S;\quad S + C = 1$$

So, $Y = CR + GST_d - CN;$

$$E = SR - GST_d + CN$$

Disturbance

To minimize disturbance, the sensitivity, S, should be small.

Because, $S = \dfrac{1}{1 + L}$, the L should be large, and since $L = GG_c$, G_c should be large for a given G. So, we need to design a controller with a large gain, G_c, over the required range of frequencies. A typical frequency for disturbance is at a low-frequency range.

Noise

To minimize the noise, the complementary sensitivity function, C, should be small. Because, $C = \dfrac{L}{1+L}$, the L should be small, and since $L = GG_c$, the G_c should be small over the required range of frequencies. So, the controller gain should be large to reject the disturbance, but smaller to attenuate the noise. This is difficult to achieve over the same frequency range. However, the typical frequency range for disturbances is low and for noise, it is high. Therefore, the control system can be designed such that the loop gain, L, is large at low frequencies (to minimize the disturbance) and small at high frequencies (to minimize the noise).

4.4. Sensitivity Analysis

$$Y(s) = \left(\frac{L}{1+L}\right)R + \left(\frac{G}{1+L}\right)T_d - \left(\frac{L}{1+L}\right)N$$

$$\text{For } T_d = 0 = N, \quad Y = \frac{L}{1+L}R$$

$$\text{If } L \gg 1, \ 1+L \approx L, \ Y \approx R$$

$$E(s) = \left(\frac{1}{1+L}\right)R - \left(\frac{G}{1+L}\right)T_d + \left(\frac{L}{1+L}\right)N$$

$$\text{For } T_d = 0 = N, \ E = \left(\frac{1}{1+L}\right)R$$

$$\text{If } L \gg 1, \ E \approx \left(\frac{1}{L}\right)R$$

Because $L = GG_c$ and if GG_c is large, any small change in plant TF, G, does not affect sensitivity.

System Sensitivity

System transfer function: $T(s) = \dfrac{Y(s)}{R(s)}$

Sensitivity of T with respect to G, the plant transfer function:

$$S_G^T = \frac{\partial T / T}{\partial G / G} = \left(\frac{\partial T}{\partial G}\right)\frac{G}{T}; \ S_G^T = \frac{\partial(\ln T)}{\partial(\ln G)}$$

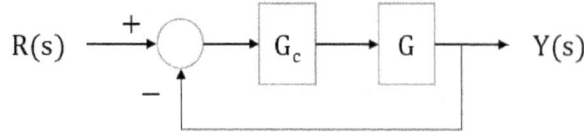

$$Y(s) = \left(\frac{G_c G}{1 + G_c G}\right) R(s)$$

$$T(s) = \frac{Y(s)}{R(s)} = \frac{G_c G}{1 + G_c G}$$

$$\frac{\partial T}{\partial G} = \frac{(1 + G_c G)G_c - G_c G(G_c)}{(1 + G_c G)^2} = \frac{G_c}{(1 + G_c G)^2}$$

$$S_G^T = \left(\frac{\partial T}{\partial G}\right)\frac{G}{T} = \frac{G_c}{(1 + G_c G)^2} \frac{G(1 + G_c G)}{G_c G} = \frac{1}{1 + G_c G} = \frac{1}{1 + L} = S, \text{ the sensitivity function}$$

If $G = f(\alpha)$, the sensitivity of T with respect to α :

$$S_\alpha^T = S_G^T S_\alpha^G = \left[\left(\frac{\partial T}{\partial G}\right)\frac{G}{T}\right]\left[\left(\frac{\partial G}{\partial \alpha}\right)\frac{\alpha}{G}\right]$$

Error and System Sensitivity

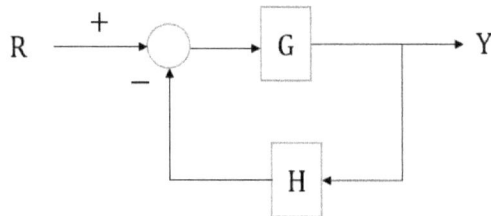

$$\text{TF: } T = \frac{Y}{R} = \frac{G}{1 + GH};$$

$$\frac{\partial T}{\partial G} = \frac{(1 + GH)(1) - G(H)}{(1 + GH)^2} = \frac{1}{(1 + GH)^2}$$

$$S_G^T = \left(\frac{\partial T}{\partial G}\right)\frac{G}{T} = \frac{1}{(1 + GH)^2} \frac{G(1 + GH)}{G} = \frac{1}{1 + GH}$$

For unity feedback: $H = 1$; $S_G^T = \dfrac{1}{1+G}$

Error: $E = R - Y = R\left(1 - \dfrac{Y}{R}\right) = R(1-T) = R\left[1 - \dfrac{G}{1+GH}\right]$

$$\frac{E}{R} = \frac{1+G(H-1)}{1+GH}$$

For $H = 1$:

$$\frac{E}{R} = \frac{1}{1+G} = S_G^T$$

EXAMPLE

Given, $G = \dfrac{K}{\tau s + 1}$; $T = \dfrac{G}{1+G}$:

Find the sensitivity of T with respect to τ.

$$S_\tau^T = S_G^T S_\tau^G$$

$$T = \frac{G}{1+G}; \qquad \frac{\partial T}{\partial G} = \frac{(1+G)-G}{(1+G)^2} = \frac{1}{(1+G)^2}$$

$$S_G^T = \left(\frac{\partial T}{\partial G}\right)\frac{G}{T} = \frac{1}{(1+G)^2}\frac{G(1+G)}{G} = \frac{1}{1+G} = \frac{\tau s + 1}{\tau s + 1 + K}$$

$$G = \frac{K}{\tau s + 1}; \quad \frac{\partial G}{\partial \tau} = -\frac{Ks}{(\tau s + 1)^2}$$

$$S_\tau^G = \left(\frac{\partial G}{\partial \tau}\right)\frac{\tau}{G} = -\frac{Ks}{(\tau s + 1)^2}\frac{\tau}{K}(\tau s + 1) = -\frac{\tau s}{(\tau s + 1)}$$

$$S_\tau^T = S_G^T S_\tau^G = \frac{(\tau s + 1)}{(\tau s + 1 + K)}\left(-\frac{\tau s}{\tau s + 1}\right) = -\frac{\tau s}{(\tau s + 1 + K)}$$

4.4.1. Disturbance Signal

An unwanted input signal affects output. For example, a car running over a speed bump experiences this action as disturbance. It is an unwanted input to the cruise control system of the car.

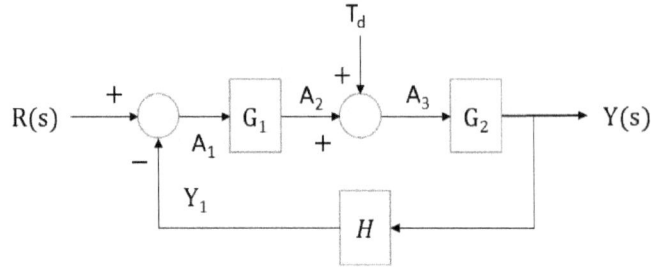

$$A_1 = R - Y_1; \ Y_1 = HY$$

$$A_1 = R - HY$$

$$A_2 = A_1 G_1 = G_1 R - G_1 HY$$

$$A_3 = A_2 + T_d = T_d + G_1 R - G_1 HY$$

$$Y = A_3 G_2 = G_2 T_d + G_1 G_2 R - G_1 G_2 HY$$

$$Y(1 + G_1 G_2 H) = G_2 T_d + G_1 G_2 R$$

$$Y = \left(\frac{G_2}{1 + G_1 G_2 H}\right) T_d + \left(\frac{G_1 G_2}{1 + G_1 G_2 H}\right) R$$

Error: $E = R - Y = R\left[1 - \dfrac{G_1 G_2}{1 + G_1 G_2 H}\right] - \left(\dfrac{G_2}{1 + G_1 G_2 H}\right) T_d$

For unity feedback ($H = 1$):

$$E = R\left(\frac{1}{1 + G_1 G_2}\right) - \left(\frac{G_2}{1 + G_1 G_2}\right) T_d$$

The error due to disturbance alone can be obtained by assuming $R = 0$:

$$(E)_d = -\left(\frac{G_2}{1 + G_1 G_2}\right) T_d$$

If $G_1 G_2 \gg 1; \ 1 + G_1 G_2 \approx G_1 G_2$

$$(E)_d = -\left(\frac{G_2}{G_1 G_2}\right) T_d = -\left(\frac{1}{G_1}\right) T_d$$

To minimize the error due to disturbance, the gain of G_1 should be large. Here, G_1 represents the controller gain. It aligns with the previous conclusion that the controller should be designed with a large gain to minimize the effect of disturbance.

4.4.2. Noise Attenuation

The error function in general:

$$E = (s)R - (Gs)T_d + (C)N$$

The error due to noise can be obtained by assuming, $R = 0 = T_d.$

Then:

$$(E)_N = (C)N = \left(\frac{L}{1+L}\right)N$$

The error due to noise is low if the loop gain, L, is low. Because $L = GG_c$, it leads to the controller gain, G_c, to be low. A small loop gain ensures good noise attenuation, whereas a large loop gain ensures rejection of disturbance. Therefore, the controller should have a high gain at low frequencies to minimize the effect of disturbance and a low gain at high frequencies to minimize the effect of noise.

EXAMPLE

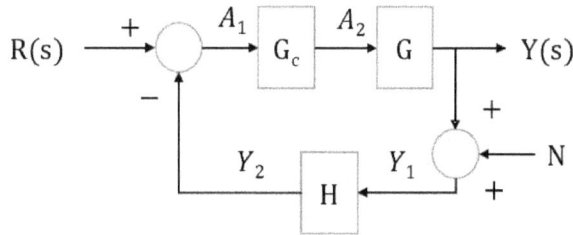

Find the error function:

$$Y_1 = Y + N$$

$$Y_2 = HY_1 = HY + HN$$

$$A_1 = R - Y_2 = R - HY - HN$$

$$A_2 = A_1 G_c = G_c R - G_c HY - G_c HN$$

$$Y = A_2 G = GG_c R - GG_c HY - GG_c HN$$

$$Y(1 + GG_cH) = GG_cR - GG_cHN$$

$$Y = \left(\frac{GG_c}{1 + GG_cH}\right)R - \left(\frac{GG_cH}{1 + GG_cH}\right)N$$

$$E = R - Y = \left[1 - \frac{GG_c}{1 + GG_cH}\right]R + \left(\frac{GG_cH}{1 + GG_cH}\right)N$$

4.4.3. Steady-State Error (e_{ss})

The steady-state error is the value of error function when the time, t, is infinity. So, the final value theorem can be applied to find the steady-state value of the error function.

$$\lim_{t \to \infty} e(t) = \lim_{s \to 0} sE(s) = e_{ss}$$

Or

$$e_{ss} = [sE(s)]_{s=0}$$

Gain setting: Controller gain can be selected such that the steady-state error is zero or equal to a given value. A study on the effect of change in gain on the steady-state error is defined as the gain setting process.

EXAMPLE

If $G(s) = \dfrac{K}{\tau s + 1}$ and $R(s) = 1/s$, find the value of K for the steady-state error to be zero.

OPEN LOOP

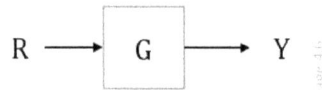

$$Y = RG$$

$$E = R - Y = R(1 - G) = \frac{1}{s}\left[1 - \frac{K}{\tau s + 1}\right]$$

$$e_{ss} = [sE(s)]_{s=0} = \left[1 - \frac{K}{\tau s + 1}\right]_{s=0} = 1 - K$$

The steady-state error, $e_{ss} = 0$ at $K = 1$.

CLOSED LOOP

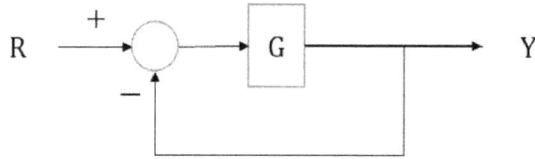

$$T(s) = \frac{Y}{R} = \frac{G}{1+G}$$

$$E(s) = R - Y = R\left(1 - \frac{Y}{R}\right) = R(1-T) = R\left[1 - \frac{G}{1+G}\right] = R\left(\frac{1}{1+G}\right)$$

$$E(s) = \frac{1}{s\left[1 + \dfrac{K}{\tau s + 1}\right]} = \frac{\tau s + 1}{s[\tau s + 1 + K]}$$

$$e_{ss} = [sE(s)]_{s=0} = \left[\frac{\tau s + 1}{\tau s + 1 + K}\right]_{S=0} = \frac{1}{1+K}$$

The steady-state error will be the minimum with a large value of K. It will be zero when the value of K is infinity.

EXAMPLE

We derived the error function for a system with noise:

$$E = R\left[1 - \frac{GG_c}{1 + GG_c H}\right] + \left(\frac{GG_c H}{1 + GG_c H}\right)N$$

If $G(S) = \dfrac{100}{S + 100}$; $G_c = K$; $H = \dfrac{1}{s+5}$; $R = \dfrac{1}{s} = N$

Determine the steady-state error for the cases (a) $N = 0$ and (b) $R = 0$.

(a) $N = 0, R = 1/s$

$$E(s) = \left[\frac{1 + GG_c H - GG_c}{1 + GG_c H}\right]\left(\frac{1}{s}\right)$$

$$GG_c = \frac{100K}{s+100}; GG_cH = \frac{100K}{(s+100)(s+5)}$$

$$1 + GG_cH - GG_c = 1 + \frac{100K}{(s+100)(s+5)} - \frac{100K}{s+100}$$

$$= \frac{(s+100)(s+5) + 100K - 100K(s+5)}{(s+100)(s+5)}$$

$$1 + GG_cH = \frac{(s+100)(s+5) + 100K}{(s+100)(s+5)}$$

$$e_{ss} = [sE(s)]_{s=0} = \left[\frac{(s+100)(s+5) + 100K - 100K(s+5)}{(s+100)(s+5) + 100K}\right]_{s=0}$$

$$e_{ss} = \frac{500 - 400K}{500 + 100K} = \frac{5 - 4K}{5 + K}$$

For $e_{ss} = 0$, $K = 5/4 = 1.25$

(b) $R = 0$, $N = 1/s$

$$E(s) = \left[\frac{GG_cH}{1 + GG_cH}\right]\left(\frac{1}{s}\right)$$

$$GG_cH = \frac{100K}{(s+100)(s+5)}$$

$$1 + GG_cH = \frac{(s+100)(s+5) + 100K}{(s+100)(s+5)}$$

$$e_{ss} = [sE(s)]_{s=0} = \left[\frac{100K}{(s+100)(s+5) + 100K}\right]_{s=0} = \frac{100K}{500 + 100K} = \frac{K}{5 + K}$$

The steady-state error will be the minimum with a lower value of controller gain.

For $K = 1$, $e_{ss} = 1/6$

EXAMPLE

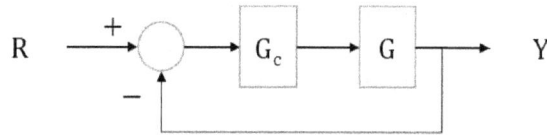

Given that $G_c = K$, and $G(s) = \dfrac{10}{s(s+2)}$:

Determine the controller gain, K (or gain setting), for the ramp input so that the steady-state error is 1%.

For ramp input: $(s) = \dfrac{1}{s^2}$; steady-state error, $e_{ss} = 0.01$.

System TF: $T(s) = \dfrac{Y}{R} = \dfrac{GG_c}{1+GG_c}$

Error function:

$$E = R(1-T) = \dfrac{1}{s^2}\left[1 - \dfrac{GG_c}{1+GG_c}\right] = \dfrac{1}{s^2}\left[\dfrac{1}{1+GG_c}\right]$$

$$1 + GG_c = 1 + \dfrac{10K}{s(s+2)} = \dfrac{s(s+2)+10K}{s(s+2)}$$

$$E(s) = \dfrac{1}{s^2}\left[\dfrac{s(s+2)}{s(s+2)+10K}\right] = \dfrac{1}{s}\left[\dfrac{s+2}{s(s+2)+10K}\right]$$

$$e_{ss} = [sE(s)]_{s=0} = \left[\dfrac{s+2}{s(s+2)+10K}\right]_{s=0} = \dfrac{2}{10K}$$

$$0.01 = \dfrac{1}{5K}, \quad K = 20$$

4.5. Time Constant

The time constant is defined as the time taken for the output to change due to the applied step input.

If the first order system *TF* is expressed in standard format as $G(s) = \dfrac{1}{\tau_c s + 1}$, then the coefficient of s, τ_c, is the time constant.

Let $\dfrac{Y}{R} = G = \dfrac{1}{\tau_c s + 1}$; $R = \dfrac{1}{s}$:

$$Y = \frac{1}{s(\tau_c s + 1)} = \frac{A}{s} + \frac{B}{\tau_c s + 1}; A = 1, B = -\tau_c$$

$$= \frac{1}{s} - \frac{\tau_c}{\tau_c s + 1}$$

$$Y(s) = \frac{1}{s} - \frac{1}{s + \dfrac{1}{\tau_c}}$$

$$y(t) = 1 - e^{-t/\tau_c} = \begin{cases} 1 \ @ \ \tau_c = 0 \\ 0 \ @ \tau_c = \infty \end{cases}$$

EXAMPLE

Open Loop

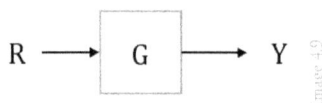

Let $G(s) = \dfrac{K}{as + b}$; $T = \dfrac{Y}{R} = G = \dfrac{K}{as + b}$.

In standard format: $T = \dfrac{K}{b}\left[\dfrac{1}{(a/b)s + 1}\right] = \dfrac{K_1}{\tau_c s + 1}$

Time constant: $\tau_c = \dfrac{a}{b}$; $K_1 = \dfrac{K}{b}$

Closed Loop

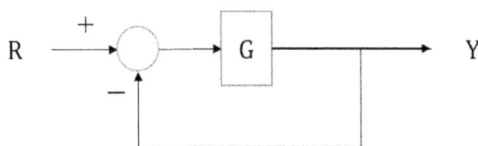

$$T = \frac{Y}{R} = \frac{G}{1+G} = \left(\frac{K}{as+b}\right) \Big/ \left[1 + \frac{K}{as+b}\right]$$

$$T(s) = \frac{K}{as+(b+K)} = \frac{K}{(b+K)} \left| \frac{1}{\left(a/(b+K)\right)s+1} \right| = \frac{K_1}{\tau_c s + 1}$$

Time constant: $\tau_c = a/(b+K);\ K_1 = K/(b+K)$

Let $K = 150,\ a = 5,\ b = 10;\ G(s) = 150/(5s+10)$

For open loop: $\tau_c = a/b = 0.5$ sec; $K_1 = 15$

For closed loop: $\tau_c = a/(b+K) = 1/32 = 0.03$ sec; $K_1 = 15/16$

4.6. Biomedical System Control

When patients are being treated, blood pressure control is important during anesthesia to regulate the depth of the anesthesia. If it is low, the patient will feel greater pain, and if it is high, the patient will die or become comatose. To measure the depth of anesthesia, the parameter generally used is the mean arterial pressure (MAP). The level of MAP serves as a guide for the delivery of anesthesia.

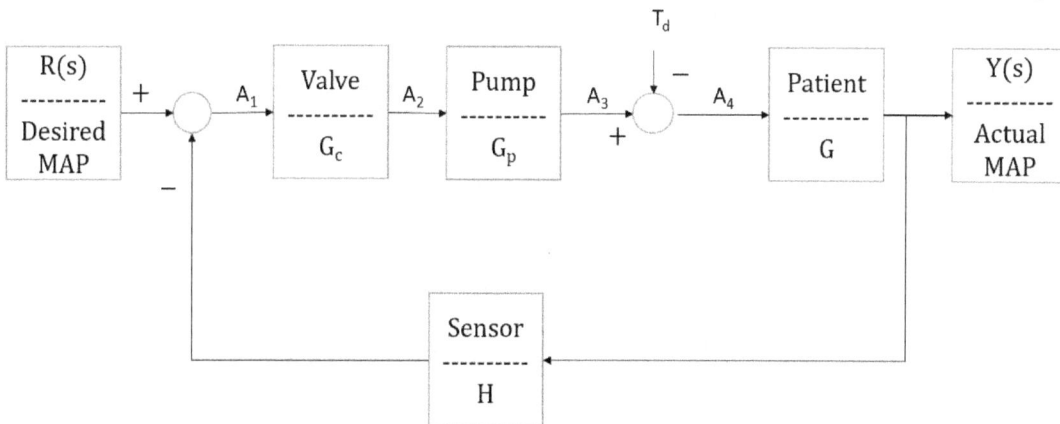

The control system is given below:

Let $G_c(s) = \dfrac{1}{s}[K_D s^2 + K_P s + K_I]$, where K_D, K_P, K_I are PID controller gains (given in Chapter 7)

Pump transfer function: $G_p(s) = \dfrac{1}{s}$; $G(s) = \dfrac{1}{(s+p)^2}$ where $p = MAP$; $H(s) = 1$

$$R(s) = R_0/s \text{ (step input)}; \quad T_d(s) = D_0/s \text{ (step disturbance)}$$

$$A_1 = R - Y, \text{ because } H = 1$$

$$A_2 = G_c A_1 = G_c R - G_c Y$$

$$A_3 = G_p A_2 = G_p G_c R - G_p G_c Y$$

$$A_4 = -T_d + A_3 = -T_d + G_p G_c R - G_p G_c Y$$

$$Y = GA_4 = -GT_d + GG_p G_c R - GG_p G_c Y$$

$$Y(1 + GG_p G_c) = (GG_p G_c)R - GT_d$$

$$Y(s) = \left(\frac{GG_p G_c}{1 + GG_p G_c}\right)R - \left(\frac{G}{1 + GG_p G_c}\right)T_d$$

$$E(s) = R(s) - Y(s) = \left(\frac{1}{1 + GG_p G_c}\right)R + \left(\frac{G}{1 + GG_p G_c}\right)T_d$$

Steady-State Error When $T_d = 0$

$$GG_p G_c = \frac{K_D s^2 + K_P s + K_I}{s^2(s+p)^2} = \frac{K_D s^2 + K_P s + K_I}{s^4 + 2ps^3 + p^2 s^2}$$

$$E(s) = \left(\frac{1}{1 + GG_p G_c}\right)\left(\frac{R_0}{s}\right) = \left(\frac{s^4 + 2ps^3 + p^2 s^2}{s^4 + 2ps^3 + (p^2 + K_D)s^2 + K_P s + K_I}\right)\left(\frac{R_0}{s}\right)$$

$$e_{ss} = [sE(s)]_{s=0} = \frac{R_0(s^4 + 2ps^3 + p^2 s^2)}{s^4 + 2ps^3 + (p^2 + K_D)s^2 + K_P s + K_I} = 0$$

The steady-state error due to the step input of magnitude, R_0, is zero.

Steady-State Output due to Step Disturbance $(R = 0)$

$$T_d = \frac{D_0}{s}$$

$$Y(s) = -\left(\frac{G}{1 + GG_pG_c}\right)T_d = \frac{-s^2}{[s^4 + 2ps^3 + (p^2 + K_D)s^2 + K_Ps + K_I]}\left(\frac{D_0}{s}\right)$$

$$\lim_{t \to \infty} y(t) = \lim_{s \to 0}[sY(s)] = \frac{-D_0s^2}{[s^4 + 2ps^3 + (p^2 + K_D)s^2 + K_Ps + K_I]} = 0$$

The step disturbance (surgical disturbance) of magnitude, D_0, does not affect the steady-state output.

Sensitivity of Transfer Function With Respect to p $(T_d = 0)$

$$S_p^T = S_G^T S_p^G$$

$$T = \frac{Y}{R} = \frac{GG_pG_c}{1 + GG_pG_c}$$

$$\frac{\partial T}{\partial G} = \frac{[(1 + GG_pG_c)(G_pG_c) - (GG_pG_c)(G_pG_c)]}{(1 + GG_pG_c)^2} = \frac{G_pG_c}{(1 + GG_pG_c)^2}$$

$$S_G^T = \left(\frac{\partial T}{\partial G}\right)\left(\frac{G}{T}\right) = \frac{G_pG_c}{(1 + GG_pG_c)^2}\frac{(1 + GG_pG_c)}{G_pG_c} = \frac{1}{1 + GG_pG_c}$$

$$G = \frac{1}{(s + p)^2}$$

$$\frac{\partial G}{\partial p} = \frac{(s + p)^2(0) - 2(s + p)}{(s + p)^4} = \frac{-2}{(s + p)^3}$$

$$S_p^G = \left(\frac{\partial G}{\partial P}\right)\left(\frac{p}{G}\right) = \frac{-2}{(s + p)^3}p(s + p)^2 = -\frac{2p}{s + p}$$

$$S_G^T = \frac{1}{1 + GG_pG_c} = \frac{s^2(s+p)^2}{[s^4 + 2ps^3 + (p^2 + K_D)s^2 + K_Ps + K_I]}$$

$$S_p^T = S_G^T S_p^G = \frac{-2ps^2(s+p)}{[s^4 + 2ps^3 + (p^2 + K_D)s^2 + K_Ps + K_I]}$$

Let $K_P = 6, K_D = 4, K_I = 1, p = 2$

$$S_p^T = \frac{(-4s^3 - 8s^2)}{(s^4 + 4s^3 + 8s^2 + 6s + 1)}$$

4.7. Aerospace System Control

In the remote exploration of the Moon's surface, the goal is to operate the robot with minimum disturbance from rocks, with a low sensitivity to changes in controller gain, K.

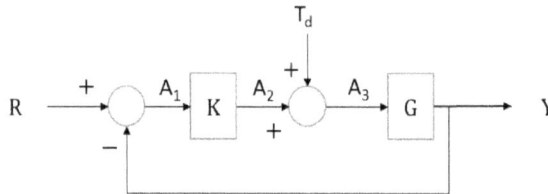

$$G(s) = \frac{1}{(s+1)(s+3)} = \frac{1}{(s^2 + 4s + 3)}$$

$$A_1 = R - Y$$

$$A_2 = A_1 K = KR - KY$$

$$A_3 = T_d + A_2 = T_d + KR - KY$$

$$Y = GA_3 = GT_d + GKR - GKY$$

$$Y(1 + GK) = (GK)R + GT_d$$

$$Y(s) = \left(\frac{GK}{1 + GK}\right)R + \left(\frac{G}{1 + GK}\right)T_d$$

Sensitivity of T(s) When $T_d = 0$

$$T = \frac{Y}{R} = \frac{GK}{1+GK}$$

$$S_K^T = \left(\frac{\partial T}{\partial K}\right)\left(\frac{K}{T}\right)$$

$$\frac{\partial T}{\partial K} = \frac{\left[(1+GK)G - GK(G)\right]}{(1+GK)^2} = \frac{G}{(1+GK)^2}$$

$$S_K^T = \frac{G}{(1+GK)^2}\frac{K(1+GK)}{GK} = \frac{1}{1+GK} = \frac{s^2+4s+3}{s^2+4s+(3+K)}$$

Let $s = j\omega$:

$$S_K^T = \frac{-\omega^2 + 4j\omega + 3}{-\omega^2 + 4j\omega + 3 + K}$$

$$= \frac{(3-\omega^2)+j(4\omega)}{(K+3-\omega^2)+j(4\omega)}$$

$$\left|S_K^T\right| = \frac{\sqrt{\left(3-\omega^2\right)^2 + 16\omega^2}}{\sqrt{\left(K+3-\omega^2\right)^2 + 16\omega^2}}$$

At low frequencies, ω^2 is negligibly small.

Therefore, $\left|S_K^T\right| = \dfrac{3}{K+3}$

For $K = 2$, $\left|S_K^T\right| = \dfrac{3}{5} = 0.6$

For, $\omega = 1$; $\left|S_K^T\right| = \dfrac{\sqrt{2^2+16}}{\sqrt{4^2+16}} = \sqrt{\dfrac{20}{32}} = 0.79$

For $0 \le \omega \le 1$ the sensitivity varies from 0.6 to 0.8.

OR, for unity feedback ($H = 1$), $\dfrac{E}{R} = [1-T] = \dfrac{1}{1+GK} = S_K^T = 1/(1+L); L = KG$.

$E = R[1/(1 + GK)] = (1/s) [1/(1 + GK)]$ for step input.

$$e_{ss} = [sE(s)]_{s=0} = \frac{s^2 + 4s + 3}{s^2 + 4s + (3 + K)} = \frac{3}{K + 3} = |S_K^T|$$

NOTE: For non-unity feedback ($H \neq 1$), $S_K^T = 1/(1 + LH)$; $L = KG$; $T = KG/(1 + KGH)$

$$E/R = [1 - T] = [1 + KG (H - 1)] / [1 + KGH] \neq S_K^T$$

Effect of Disturbance (R = 0)

$$Y = \left(\frac{G}{1 + GK}\right) T_d; \text{ let } T_d = \frac{1}{s}$$

$$Y = \left[\frac{1}{s^2 + 4s + (3 + K)}\right] \frac{1}{s}$$

Steady-state response: $\lim_{t \to \infty} y(t) = \lim_{s \to 0} sY(s)$

$$y(t) = \left[\frac{1}{s^2 + 4s + 3 + K}\right]_{s=0} = \frac{1}{3 + K}$$

For $K = 2$, $y(t) = 1/5 = 0.2$.

To minimize the effect of disturbance, choose a higher value for K.

Let $K = 100$; $y(t) = 1/103 = 0.0097$.

The effect of disturbance is negligibly small on the steady-state response with a larger value of K.

4.8. Summary

In this chapter, a method was presented to study the error analysis with disturbance and noise. The forward-loop gain ($L = GG$) should be high at low frequencies to minimize the effect of disturbance, whereas the loop gain should be low at high frequencies to minimize the effect of noise. The final value theorem is useful for determining the steady-state error of a given system. A time constant is introduced to find the time taken for the output to change due to an applied step input. The example of blood pressure control provided a typical example regarding biomedical control systems. Similarly, a study on the robot on the Moon's surface gave an example regarding aerospace control systems.

4.9. Assessment

1. The error is given as:

    a. $E = R - Y$

    b. $E = R(1 - T)$

    c. $E = R - R(Y/R)$

    d. All of the above

2. To minimize the disturbance, the controller should have:

    a. A large gain at high frequencies

    b. A small gain at high frequencies

    c. A large gain at low frequencies

    d. A small gain at low frequencies

3. To minimize the noise, the controller should have:

    a. A small gain at high frequencies

    b. A small gain at low frequencies

    c. A large gain at low frequencies

    d. A large gain at high frequencies

4. The sensitivity of a system, TF (T), with respect to the plant TF, (G), is given by:

    a. $S_G^T = \left(\dfrac{\partial T}{\partial G}\right)\left(\dfrac{T}{G}\right)$

    b. $S_G^T = \left(\dfrac{\partial T}{\partial G}\right)\Big/\left(\dfrac{T}{G}\right)$

    c. $S_T^G = \left(\dfrac{\partial G}{\partial T}\right)\left(\dfrac{T}{G}\right)$

    d. None of the above

5. The sensitivity of $T(s)$ with respect to a parameter (α) in $G(s)$ is given by:

    a. $S_\alpha^T = \dfrac{\partial T / T}{\partial \alpha / \alpha}$

b. $S_\alpha^T = S_G^T S_\alpha^G$

c. $S_\alpha^T = \left(\dfrac{\partial T}{\partial G}\right)\left(\dfrac{G}{T}\right)\left(\dfrac{\partial G}{\partial \alpha}\right)\left(\dfrac{\alpha}{G}\right)$

d. All of the above

6. The steady-state error is the value of the error function, $e(t)$, evaluated at:

    a. $t = 0$

    b. $t = \infty$

    c. $s = \infty$

    d. All of the above

7. Gain setting is defined as the effect of:

    a. Change in gain on the error function

    b. Change in gain on the steady-state error

    c. Error on the gain

    d. Gain on sensitivity

8. If $E(s) = \dfrac{(s+4)(s+5)}{s(s^2+3s+2)}$, the steady-state error is:

    a. 10

    b. 0

    c. 0.1

    d. ∞

9. If $E(s) = \left(\dfrac{1}{s^3}\right)\left(\dfrac{s^3+2s^2}{s^3+4s^2+5s+1}\right)$, the steady-state error is:

    a. 0.5

    b. ∞

    c. 0

    d. 2

10. If $E(s) = \dfrac{s^2 + 3s + 4}{s^3 + 4s^2 + (5 + 2K)s}$ and $e_{ss} = 0.05$, the value of K is:

   a. 375

   b. 3.75

   c. 37.5

   d. 0.375

4.10. Practice Problems

1. Answer all the assessment questions in Section 4.9

2. If $G(s) = 10/(ts + 1)$, determine (a) the error for given impulse input, and (b) sensitivity of transfer function with respect to the parameter, t

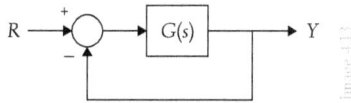

3. If $G(s) = 20/(s + 20)$ and $H(s) = 1/(s + 10)$, determine (a) Y/R at $N = 0$, and (b) Y/N at $R = 0$

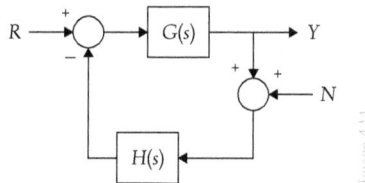

4. A system with disturbance is given below. If $G(s) = K/(s + 10)$ and $H(s) = 14/(s^2 + 5s + 6)$, determine, (a) The output function Y(s), (b) The transfer function, Y(s)/R(s) at $T_d = 0$, (c) The error function, E(s) and steady state error due to step input, (d) The steady state response due to step disturbance at $R = 0$.

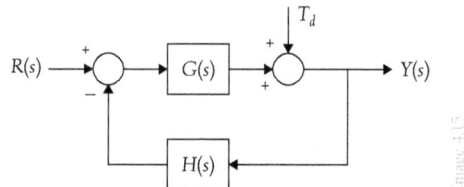

5. If $G(s) = 200/(s^2 + 25s + 200)$, and $G_c(s) = K/(0.1 s + 1)$ for a climate control system given below, determine (a) The output function Y(s), (b) The sensitivity of the system transfer function $T(s)$ with respect to K at $T_d = 0$, and (c) The effect of disturbance on the output, $Y(s)/T_d(s)$ at $R = 0$.

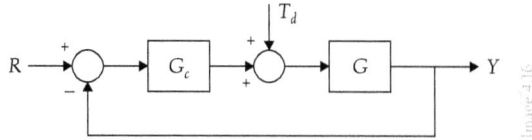

6. The block diagram of a machine tool control system is given below. If $G_c = K$, and $G(s) = b/(s + 2)$ determine (a) The output function $Y(s)$, (b) Transfer function $T(s)$ at $T_d = 0$, (c) The sensitivity of transfer function with respect to the parameter "b".

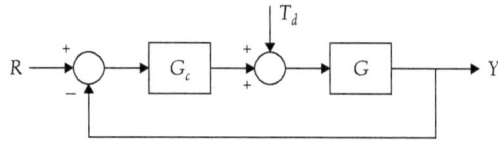

7. If $G(s) = (s + 125)/\{s(s + 5)\}$; determine the steady state error for ramp input.

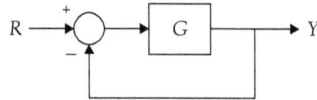

8. If $G(s) = 50/(s + 100)$; determine the value of K (controller gain setting) such that the steady state error is 0.004

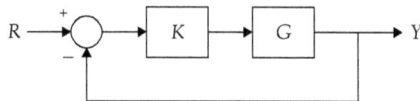

9. The error function due to noise is given as, $E(s) = [GH/(1 + GH)]\{N(s)\}$

   If $G(s) = K/(s + 19)$, $H(s) = 1/(s + 5)$, and $N(s) = 1/s$ determine the value of K such that the steady state error is five percent.

10. If $G(s) = 180/(3s + 20)$ determine the transfer function. If it can be expressed in the format of $T(s) = K_1/(\tau_c S + 1)$; determine the value of K_1 and time constant, τ_c

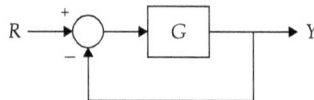

5

PERFORMANCE OF CONTROL SYSTEMS

5.1. Introduction

In this chapter, emphasis is placed on identifying the performance parameters in order to study a control system's transient and steady-state responses. Performance parameters are defined based on the step response in the time domain. Rise time, peak time, settling time, final value, and percent overshoot are used as typical performance parameters. Steady-state errors are computed for type 0, type 1, and type 2 systems with step input, ramp input, and quadratic input. A mechanical system is given as typical example to study the performance of control system.

5.2. Learning Objectives

1. Identify performance parameters for the control of transient and steady-state responses of a control system.

2. Apply time domain performance specifications to a second-order system response.

3. Study the steady-state error for step, ramp, and quadratic input functions.

In general, a control system represents a dynamic system, and hence its performance is given in terms of its transient and steady-state responses. Performance parameters serve as measures to study how well a control system performs and how to change the parameters for a desired performance.

5.3. Transient and Steady-State Responses

The steady-state response and steady-state error are derived for a first-order system and a second-order system. The final value theorem is used in this process.

5.3.1. First-Order Systems

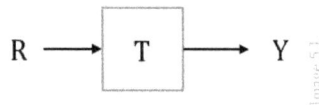

$$R \longrightarrow \boxed{T} \longrightarrow Y$$

Image 5.1

Let $T(s) = \dfrac{9}{s+10}$ = transfer function (TF) for a first-order system:

$$R(s) = 1/s \text{ (step input)}$$

$$Y(s) = T(s)R(s) = \frac{9}{s(s+10)} = \frac{0.9}{s} - \frac{0.9}{s+10} \text{ (by partial fraction)}$$

Taking an inverse Laplace transform (LT):

$$y(t) = 0.9(1 - e^{-10t}) = \text{transient response}$$

The steady-state response (at t, equal to infinity):

$$y(\infty) = 0.9(1 - 0) = 0.9$$

Also, by the final value theorem:

$$\lim_{t \to \infty} Y(t) = \lim_{s \to 0}[sY(s)]; \ Y(s) = \frac{9}{s(s + 10)}$$

$$Y(\infty) = [sY(s)]_{s=0} = \left[\frac{9}{s + 10}\right]_{s=0} = 9/10 = 0.9$$

Error: $E = R(1 - T)$

$$E(s) = \frac{1}{s}\left[1 - \frac{9}{s + 10}\right] = \frac{1}{s}\left(\frac{s + 1}{s + 10}\right)$$

Steady-state error: $e_{ss} = [sE(s)]_{s=0} = \left(\frac{s + 1}{s + 10}\right)_{s=0} = 1/10 = 0.1$

$$= [1 - y(\infty)] = 1 - (0.9) = 0.1$$

5.3.2. Second-Order Systems

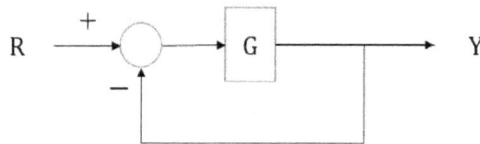

$$T(s) = \frac{G}{1 + G}$$

$$\text{Let } G(s) = \frac{\omega_n^2}{s^2 + \left(2\zeta\omega_n\right)s} = \text{ TF of a second-order system}$$

$$T(s) = \frac{\omega_n^2}{s^2 + \left(2\zeta\omega_n\right)s + \omega_n^2}; \ \zeta < 1(\text{underdamped})$$

Let $R(s) = 1/s$ (step input):

$$Y(s) = T(s)R(s)$$

$$= \frac{\omega_n^2}{s\left[s^2 + \left(2\zeta\omega_n\right)s + \omega_n^2\right]}$$

Taking the inverse LT (from the LT table):

$$y(t) = 1 - \frac{e^{-\zeta\omega_n t}}{\sqrt{1-\zeta^2}}\sin\left(\omega_d t + \phi\right)$$

$$\phi = \cos^{-1}\zeta;\ \omega_d = \omega_n\sqrt{1-\zeta^2}$$

Let $R(s) = 1$ (impulse input):

$$Y(s) = \frac{\omega_n^2}{s^2 + \left(2\zeta\omega_n\right)s + \omega_n^2} = s[Y(s)]_{\text{unit step}}$$

$$y(t) = \frac{\omega_n}{\sqrt{1-\zeta^2}}e^{-\zeta\omega_n t}\sin\omega_d t = \frac{d}{dt}[y(t)]_{\text{unit step}}$$

Final value:

$$\text{Step input: } y(\infty) = [sY(s)]_{s=0} = \frac{\omega_n^2}{\omega_n^2} = 1$$

$$\text{Impulse: } y(\infty) = [sY(s)]_{s=0} = \left[s\omega_n^2/\left(s^2 + 2\zeta\omega_n s + \omega_n^2\right)\right]_{s=0} = 0$$

5.4. Performance Parameters

Performance parameters are defined in terms of the step response of the closed-loop system. Typical parameters used for performance measure are as follows:

   i. Rise time

  ii. Peak time

 iii. Settling time

  iv. Final value

   v. Percent overshoot

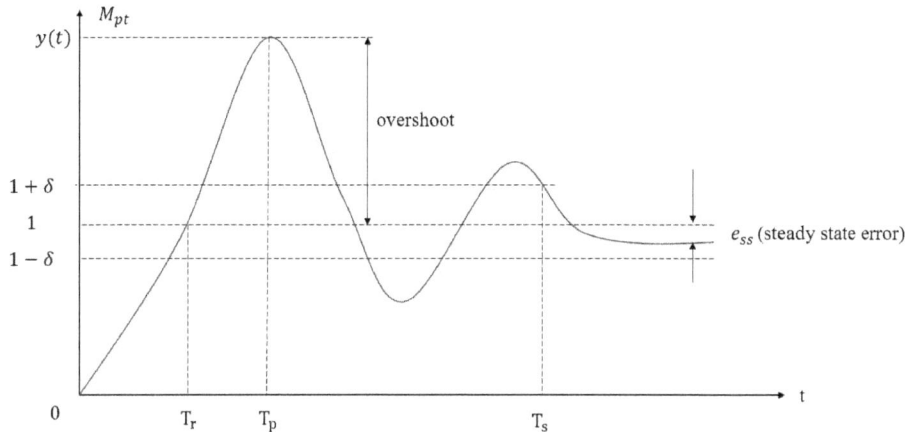

The unit step response is given below.

*T_r: rise time (measures swiftness of response)

*T_p: peak time (the time at peak value of response)

*T_s: settling time (time required to settle with in some $\delta\%$ of final value)

M_{pt}: peak value at $t = T_p$, $(M_{pt} = y(T_p))$; $M_{pt} = 1 + overshoot$

*FV: final value $->$ steady-state value at $t = \infty \left[y(\infty) = sY(s)_{s=0}\right] = 1$ for step input in the figure above

*PO: percent overshoot $->$ difference between the peak value and final value expressed as a percentage with respect to the final value.

$$P.O = \left|\frac{M_{pt} - FV}{FV}\right|100$$

δ: percent of FV used as a reference to find T_s (typical value of δ is equal to 2% of FV; for FV $= 1, \delta = 2\% = 0.02$).

The exponential decay in the response is $e^{-\zeta\omega_n t}$.

$$@ t = T_s, e^{-\zeta\omega_n T_s} = \delta = 0.02$$

$$\text{i.e.}, \frac{1}{e^{\zeta\omega_n T_s}} = 0.02 \Rightarrow e^{\zeta\omega_n T_s} = 1/0.02 = 50$$

$$\zeta\omega_n T_s = \ln 50 = 3.91 \approx 4$$

$$T_s = \frac{4}{\zeta\omega_n}; \text{ let the time constant} = \frac{1}{\zeta\omega_n}; \text{ then } T_s = 4\tau$$

For an arbitrary value of δ, $e^{-\zeta\omega_n T_s} = \delta$:

$$-\zeta\omega_n T_s = \ln\delta$$

$$T_s = -\frac{\ln\delta}{\zeta\omega_n}$$

For $\delta = 2\%$; $\ln(0.02) = -4$; $T_s = \frac{4}{\zeta\omega_n}$

For $\delta = 5\%$; $\ln(0.05) = -3$; $T_s = \frac{3}{\zeta\omega_n}$

5.4.1. Transient Performance Measures

Swiftness: rise time, T_r, and peak time, T_p

Closeness to desired response: settling time, T_s, and PO

To find the peak value, M_{pt}:

$$M_{pt} = y\left(T_p\right) = y_{max}$$

$$y(t)\big|_{step} = 1 - \frac{1}{\beta}e^{-\zeta\omega_n t}\sin\left(\omega_d t + \phi\right); \; \beta = \sqrt{1-\zeta^2}; \; \omega_d = \omega_n \beta$$

To find y_{max}, $\frac{d}{dt}\left(y_{step}\right) = 0$

Because $\frac{d}{dt}\left[y_{step}\right] = y\big|_{imp}$

For y_{max}, $y\big|_{imp} = 0$; $y\big|_{imp} = \frac{\omega_n}{\beta}e^{-\zeta\omega_n t}\sin\left(\omega_d t\right) = 0 \Rightarrow \sin\left(\omega_d t\right) = 0$

$$\omega_d t = n\pi; \text{ for } n = 1; t = \frac{\pi}{\omega_d} = T_p$$

So $M_{pt} = 1 - \frac{1}{\beta}e^{-\zeta\omega_n T_p}\sin\left(\omega_d T_p + \phi\right)$

$$\omega_d T_p = \pi, \; \omega_n T_p = \pi/\beta$$

$$M_{pt} = 1 - \frac{1}{\beta}e^{-\left(\frac{\zeta\pi}{\beta}\right)}\sin\left(\pi + \phi\right)$$

$$= 1 + e^{-\left(\frac{\zeta\pi}{\beta}\right)} \frac{\sin\phi}{\beta}$$

$$= 1 + e^{-\left(\frac{\zeta\pi}{\beta}\right)} = 1 + \text{overshoot}$$

$$\zeta = \cos\phi; \ \beta = \sqrt{1 - \zeta^2} = \sin\phi$$

$$\text{overshoot} = e^{-\left(\frac{\zeta\pi}{\beta}\right)}$$

$$PO = 100e^{-\left(\frac{\zeta\pi}{\beta}\right)}$$

Also, from the definition of PO:

$$PO = \left|\left(\frac{M_{pt} - FV}{FV}\right)\right| 100$$

For $FV = 1$, $PO = 100\left(M_{pt} - 1\right) = 100e^{-\left(\frac{\zeta\pi}{\beta}\right)}$.

To summarize, the performance parameters are:

1. Settling time: $T_s = \dfrac{4}{\zeta\omega_n} = 4\tau; \ \tau = \dfrac{1}{\zeta\omega_n} = \text{time constant}$

2. PO: $100e^{-\left(\frac{\zeta\pi}{\beta}\right)}; \ \beta = \sqrt{1 - \zeta^2}$

3. Peak time: $T_p = \pi/\omega_d; \ \omega_d = \omega_n\sqrt{1 - \zeta^2}$

4. Rise time: $T_r = \dfrac{1}{\omega_n}[2.16\zeta + 0.6]; \ 0.3 \leq \zeta \leq 0.8$. Rise time is defined as the swiftness of step response to rise from 10% to 90% of the magnitude of the step input. A linear approximation given above is commonly used to determine the rise time.

5. Peak value:

$$M_{pt} = 1 + e^{-\left(\frac{\zeta\pi}{\beta}\right)} = 1 + \frac{P.O.}{100}$$

6. Final value: $FV = [SY(s)]_{s=0} = 1$, for step input

The PO should be lower for closeness to the desired response. The peak time should be low for swiftness to reach the peak value. With an increase in the damping ratio, the PO decreases whereas the peak time increases.

At $\zeta = 0$, PO = 100 and $T_p = \pi/\omega_n$; At $\zeta = 1$, PO = 0 and $T_p = \infty$

The settling time decreases with an increase in the damping ratio. The rise time increases with an increase in the damping ratio. The peak value is directly proportional to the PO.

All of these performance parameters are functions of natural frequency and the damping ratio. With second-order systems, the values of w_n and ζ can be determined from their TF. Then, all the performance parameters can be calculated.

Similarly, w_n and ζ can be evaluated from the given values of the PO and T_s:

$$e^{-\left(\frac{\zeta\pi}{\beta}\right)} = \frac{P.O.}{100}; \ e^{\left(\frac{\zeta\pi}{\beta}\right)} = \frac{100}{P.O.}$$

$$\frac{\zeta\pi}{\beta} = \ln\left[\frac{100}{P.O.}\right] = \alpha$$

$$\frac{\zeta\pi}{\sqrt{1-\zeta^2}} = \alpha$$

$$\zeta^2\pi^2 = \alpha^2\left(1-\zeta^2\right)$$

$$\zeta^2(\alpha^2 + \pi^2) = \alpha^2$$

$$\zeta = \alpha/\left(\sqrt{\alpha^2 + \pi^2}\right); \ w_n = 4/\left(\zeta T_s\right)$$

Then, the other parameters, M_{pt}, T_p, T_r, can be easily calculated.

EXAMPLE

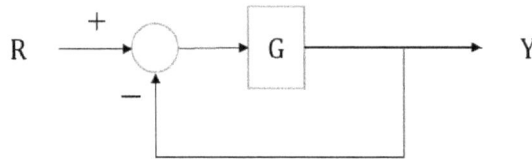

Let $G = \dfrac{K}{S(S+p)}$.

Select the gain, K, and pole, p,

such that PO = 4.3% and T_s = 4 sec.

Let us first find w_n and ζ.

$$\alpha = \ln\frac{100}{PO} = \ln\frac{100}{4.3} = 3.146$$

$$\zeta = \alpha / \left(\sqrt{\alpha^2 + \pi^2} \right) = 3.146 / \sqrt{3.146^2 + \pi^2} = 0.707$$

$$\omega_n = 4 / \zeta T_s = 4 / [0.707 \times 4] = 1.414$$

System TF: $T = \dfrac{G}{1+G} = \dfrac{K}{S(S+p)+K} = \dfrac{K}{S^2 + pS + K}$

Comparing this with a standard TF for a second-order system:

$$T = \frac{\omega_n^2}{s^2 + \left(2\zeta\omega_n\right)s + \omega_n^2}$$

$$p = 2\zeta\omega_n = 2(0.707)(1.414) = 2$$

$$K = \omega_n^2 = (1.414)^2 = 2$$

5.4.2. Steady-State Error

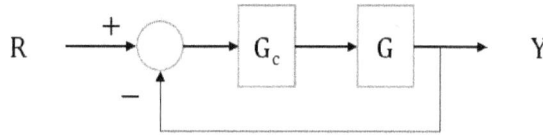

TF: $T(s) = \dfrac{Y}{R} = \dfrac{G_c G}{1 + G_c G} = \dfrac{L}{1+L}$

Error: $E(s) = R(1 - T)$

$$= R(s)\left(\frac{1}{1+L}\right)$$

Steady-state error: $e_{ss} = [SE(s)]_{s=0}$

Let us study e_{ss} for different input functions.

Test input functions:

Step: $r(t) = A; \ R(s) = A/s$

Ramp: $r(t) = (At); \ R(s) = A/s^2$

Quadratic: $r(t) = (At^2/2); \ R(s) = A/s^3$

5.4.2.1 STEP INPUT

$R(s) = A/s$, where A is the magnitude of step function:

$$E(s) = \frac{A}{s}\left(\frac{1}{1 + L(s)}\right) = \frac{A}{s}\left(\frac{1}{1 + G_c(s)G(s)}\right)$$

$$e_{ss} = \frac{A}{1 + G_c(0)G(0)}$$

In general:

$$G_c(s)G(s) = \frac{K\prod_{i=1}^{M}\left(S + Z_i\right)}{S^N\prod_{k=1}^{Q}\left(S + P_k\right)}; Z_i \rightarrow zeros \ P_k \rightarrow poles$$

$N \rightarrow$ system-type number

N	0	1	2
$1/S^N$	1	$1/S$	$1/S^2$

Type Zero $(N = 0)$

$$G_c(s)G(s) = \frac{K\prod\left(S + Z_i\right)}{\prod\left(S + P_k\right)}$$

$$G_c(0)G(0) = \frac{K\prod(Z_i)}{\prod(P_k)} = K_p = \text{position error constant}$$

Therefore:

$$e_{ss} = \frac{A}{1 + K_p}; \ K_p = \left[G_c(s)G(s)\right]_{s=0}$$

$$\text{For } N \geq 1, \ e_{ss} = \frac{A}{1 + \left[\dfrac{K\prod\left(S + Z_i\right)}{S^N\prod\left(S + P_k\right)}\right]_{s=0}}$$

$$= \left[\frac{AS^N\prod\left(S + P_k\right)}{S^N\prod\left(S + P_k\right) + K\prod\left(S + Z_i\right)}\right]_{s=0} = 0$$

The steady-state error is zero for all systems except for type 0 systems.

Example

Let $R(s) = 1/s$ $A = 1$, and let $L(s) = G_c(s)G(s) = \dfrac{4\,(s+2)}{(s+3)(s+4)}$; $K = 4$:

$$e_{ss} = \frac{A}{1+K_p}; \quad Kp = L(0) = \frac{4\,(2)}{(3)(4)} = \frac{2}{3}; \quad e_{ss} = \frac{3}{5}$$

OR: $E(s) = \dfrac{A}{s}\left(\dfrac{1}{1+L(s)}\right) = \left(\dfrac{1}{s}\right)\left[\dfrac{(s+3)(s+4)}{(s+3)(s+4)+4(s+2)}\right]$

Steady-state error: $e_{ss} = \Big[SE(s)\Big]_{s\,=\,0} = \dfrac{12}{20} = \dfrac{3}{5}$

5.4.2.2 RAMP INPUT

$$r(t) = At; \; R(s) = A/s^2$$

$$E(s) = \left(\frac{1}{1+G_cG}\right)\left(\frac{A}{S^2}\right)$$

$$e_{ss} = \Big[SE(s)\Big]_{s\,=\,0} = \frac{A}{S\big[1+G_c(s)G(s)\big]}\Big|_{s\,=\,0}$$

If $G_cG \gg 1$, $1+G_cG \approx G_cG$; $e_{ss} = \dfrac{A}{SG_c(s)G(s)}\Big|_{s\,=\,0}$

Type 1 Systems (N = 1)

$$G_cG = \frac{K\Pi\big(S+Z_i\big)}{S\Pi\big(S+P_k\big)}$$

$$e_{ss} = \frac{A\Pi\big(S+P_k\big)}{K\Pi\big(S+Z_i\big)}\Big|_{s\,=\,0} = \frac{A\Pi\big(P_k\big)}{K\Pi\big(Z_i\big)} = \frac{A}{K_v}$$

$$K_v = K\Pi\big(Z_i\big)/\Pi\big(P_k\big) = \Big[SG_c(s)G(s)\Big]_{s\,=\,0} = \text{velocity error constant}$$

For $N > 1$, $e_{ss} = 0$ and for $N < 1$, $e_{ss} = \infty$.

For the ramp input, the steady-state error is finite only for type 1 systems.

Example

Let $R(s) = \dfrac{1}{S^2}$, $A = 1$, and let $L(s) = G_c(s)G(s) = \dfrac{5\,(s+2)}{s(s+3)(s+4)}$; $K = 5$:

$$e_{ss} = \frac{A}{K_v}; \quad K_v = \left[SG_c(s)G(s) \right]_{s=0} = \frac{10}{12}; \quad e_{ss} = 1.2$$

OR:

$$E(s) = \left(\frac{A}{S^2} \right)\left(\frac{1}{1+L(s)} \right) = \left(\frac{1}{S^2} \right)\left[\frac{s(s+3)(s+4)}{s(s+3)(s+4) + 5(s+2)} \right]$$

$$= \left(\frac{1}{s} \right)\left[\frac{(s+3)(s+4)}{s(s+3)(s+4) + 5(s+2)} \right]$$

Steady-state error: $e_{ss} = \left[SE(s) \right]_{s=0} = \dfrac{12}{10} = 1.2$

5.4.2.3 QUADRATIC INPUT

$$r(t) = \frac{At^2}{2}; \quad R(s) = \frac{A}{2}\left(\frac{2}{S^3} \right) = \frac{A}{S^3}$$

$$E(s) = \left(\frac{1}{1+G_cG} \right)\left(\frac{A}{S^3} \right) \approx \frac{A}{S^3\left[G_c(s)G(s) \right]}$$

$$e_{ss} = \left[SE(s) \right]_{s=0} = \frac{A}{S^2\left[G_c(s)G(s) \right]}\Big|_{s=0}$$

Type 2 Systems (N = 2)

$$G_cG = \frac{K\Pi\left(S + Z_i \right)}{S^2\,\Pi\left(S + P_k \right)}$$

$$e_{ss} = \frac{A\Pi\left(S + P_k \right)}{K\Pi\left(S + Z_i \right)}\Big|_{s=0} = \frac{A\Pi\left(P_k \right)}{K\Pi\left(Z_i \right)} = \frac{A}{K_a}$$

$$K_a = K\Pi\left(Z_i \right)/\Pi\left(P_k \right) = \left[S^2 G_c(s)G(s) \right]_{s=0} = \text{acceleration error constant}$$

For N > 2, $e_{ss} = 0$ and for N < 2, $e_{ss} = \infty$.

For quadratic input, the steady-state error is finite only for type 2 systems.

Example

Let $R(s) = \dfrac{3}{S^3}$, $A = 3$, and let $L(s) = G_c(s)G(s) = \dfrac{10\,(s+2)}{S^2(s+3)(s+4)}$; $K = 10$:

$$e_{ss} = \frac{A}{K_a}; K_a = \left[S^2 G_c(s)G(s)\right]_{s=0} = \frac{20}{12}; e_{ss} = (3)\left(\frac{12}{20}\right) = \left(\frac{9}{5}\right)$$

OR:

$$E(s) = \left(\frac{A}{S^3}\right)\left(\frac{1}{1+L(s)}\right) = \left(\frac{3}{S^3}\right)\left[\frac{S^2(s+3)(s+4)}{S^2(s+3)(s+4)+10(s+2)}\right]$$

$$= \left(\frac{3}{s}\right)\left[\frac{(s+3)(s+4)}{S^2(s+3)(s+4)+10(s+2)}\right]$$

Steady-state error: $e_{ss} = \left[SE(s)\right]_{s=0} = (3)\left(\frac{12}{20}\right) = \left(\frac{9}{5}\right)$

The steady-state error for various inputs are summarized in the table below.

System type	Step input	Ramp input	Quadratic input
0	$A/(1+K_p)$	∞	∞
1	0	A/K_v	∞
2	0	0	A/K_a

Example

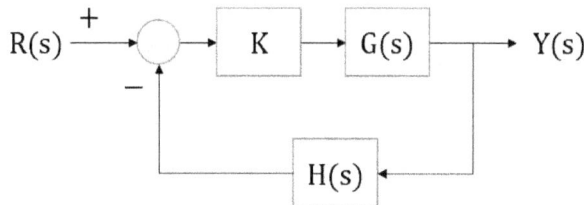

$$G(s) = \frac{1}{s+2}$$

$$H(s) = \frac{2}{s+4}$$

$$R(s) = \frac{1}{s}$$

Find K, such that the steady-state error is zero.

$$TF : T(s) = \frac{KG}{1+KGH}$$

$$E(s) = R(1-T) = \frac{1}{S}\left[\frac{1+KGH-KG}{1+KGH}\right]$$

$$e_{ss} = \left[SE(s)\right]_{s=0} = \left[\frac{1+KG(0)H(0)-KG(0)}{1+KG(0)H(0)}\right]$$

$G(0) = 1/2, H(0) = 2/4 = 1/2$

$$e_{ss} = \frac{1+K(1/4-1/2)}{1+(K/4)} = 0$$

$$1 - \frac{K}{4} = 0 \qquad K = 4$$

5.5. Mechanical System Control

The control system for a typical mechanical system is given below.

a. Choose K and K_1 so that the PO is 10% and the settling time is $(2/3)$ sec. Assume step input, $R(s)$, $=1/s$ and $T_d(s) = 0$.

b. Find the steady-state error due to the ramp input.

c. Find the effect of the step disturbance on steady-state response. Assume $R(s) = 0$.

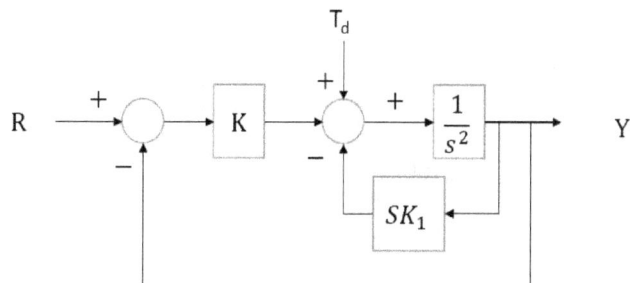

Reduced Block Diagram

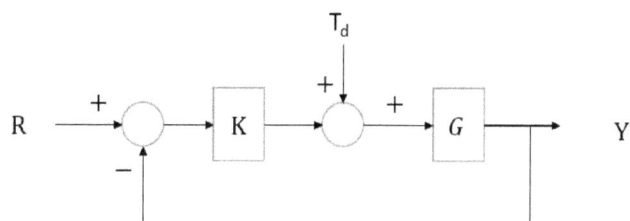

$$Y = [KG/(1+KG)]\,R + [G/(1+KG)]\,T_d$$

$$G(s) = \frac{(1/S^2)}{1+(1/S^2)(K_1 S)} = \frac{1}{S^2 + K_1 S}$$

(a) **Step input**

PO $= 10\%$; $T_s = (2/3)$ sec

$$\alpha = \ln\left(\frac{100}{PO}\right) = \ln 10 = 2.3$$

$$\zeta = \alpha\Big/\left(\sqrt{\alpha^2 + \pi^2}\right) = 0.6$$

$$\omega_n = 4\Big/\left(\zeta T_s\right) = 10$$

$$T_d = 0,\ T(s) = \frac{KG}{1+KG} = \frac{Y}{R} = \frac{K}{(S^2 + K_1 S + K)}$$

Comparing $T(s)$ with standard form of TF: $T(s) = \dfrac{\omega_n^2}{s^2 + (2\zeta\omega_n)s + \omega_n^2}$

$$K = \omega_n^2 = 100, \; K_1 = 2\zeta\omega_n = 1.2\omega_n = 12$$

(b) Ramp input

$$R(s) = \frac{1}{S^2}$$

$$E(s) = R(1 - T) = \frac{1}{S^2}\left(\frac{1}{1 + KG}\right) = \frac{1}{S^2}\left(\frac{S^2 + K_1 S}{S^2 + K_1 S + K}\right) = \frac{1}{S}\left(\frac{S + K_1}{S^2 + K_1 S + K}\right)$$

$$e_{ss} = \Big[SE(s)\Big]_{s=0} = \frac{K_1}{K} = 0.12$$

(c) Step disturbance

$$T_d = 1/S, \; R = 0$$

$$\frac{Y}{T_d} = \frac{G}{1 + KG}$$

$$Y(s) = T_d\left(\frac{1}{S^2 + K_1 S + K}\right)$$

$$= \frac{1}{S}\left(\frac{1}{S^2 + K_1 S + K}\right)$$

$$y(\infty) = SY(s)\,|_{s=0} = \frac{1}{K}$$

The effect of T_d is less for a large value of K. For $K = 100$, $y(\infty) = 0.01$.

5.6. Summary

The response of a second-order system, $Y(s)$, can be easily found from its TF, $T(s)$. The steady-state value (or final value) is one for step input and zero for impulse input. PO and settling time are the typical parameters used in the study of control system performance. The damping ratio can be determined from the PO, whereas the natural frequency can be determined from the settling time. Then the peak value, peak time, and rise time can be easily determined. Regarding steady-state error, it is shown that the error has finite value for type 0 systems with only step input, for type 1 systems with only ramp input, and for type 2 systems with only quadratic input. For the application of this study, a typical mechanical system is given as an example.

5.7. Assessment

1. $T = \dfrac{9S}{S^2 + 10S}$ represents:

   a. A first-order system

   b. A second-order system

   c. A third-order system

   d. All of the above

2. The final value theorem is useful to find the:

   a. Steady-state error

   b. Steady-state response

   c. Final value of the output

   d. All of the above

3. If $E(s) = \dfrac{8s^2 + 5s}{9s^3 + 4s^2}$, the steady-state error is:

   a. 8/9

   b. 9/8

   c. 5/4

   d. 4/5

4. If $T_s = K/\zeta\omega_n$, the value of K for a settling time within 2% of the final value is:

   a. 4.6

   b. 4.0

   c. 3.0

   d. 2.3

5. In the above problem, the value of K for a settling time within 5% of the final value is:

   a. 4.6

   b. 3.9

   c. 3.0

   d. 2.3

6. For step input, the steady-state error is zero for:

   a. Type 1 systems

   b. Type 2 systems

   c. Type 3 systems

   d. All of the above

7. For ramp input, the steady-state error is NOT zero for:

   a. Type 1 systems

   b. Type 2 systems

   c. Type 3 systems

   d. All of the above

8. For quadratic input, the steady-state error, $e_{ss} = \dfrac{A}{K_a}$ for:

   a. Type 1 systems

   b. Type 2 systems

   c. Type 3 systems

   d. All of the above

9. The error function is defined for:

   a. $E(s) = R(s) - Y(s)$

   b. $E(s) = R(s)[1 - T(s)]$

   c. $E(s) = -R(s)\left[\dfrac{Y(s) - R(s)}{R(s)}\right]$

   d. All of the above

10. If the PO and settling time are given, we can find:

    a. The damping ratio

    b. The natural frequency

    c. The peak time and peak value

    d. All of the above

5.8. Practice Problems

1. Answer all the assessment questions in Section 5.7

2. The exponential decay in terms of settling time can be expressed as $\delta = e^{-(T_s/\tau)}$, where "T_s" is the settling time and the time-constant, $\tau = \dfrac{1}{(\xi \omega_n)}$

    a. Plot the variation of settling time to time-constant ratio as a function of δ, for a range of $0.01 < \delta < 0.1$ in steps of 0.005

    b. Determine the time constant and settling time for $\delta = 8\%$ to the dynamic system described by, $5\,\ddot{y} + 48\,\dot{y} + 180\,y = 0$

3. A dynamic system is described by $5\,\ddot{y} + 20\,\dot{y} + 125\,y = 5\,u(t)$. Determine the settling time, percent overshoot, peak time, rise time, and peak value for given $\delta = 2\%$

4. Determine the time constant, peak time, rise time, and peak value for given percent overshoot of 20% and settling time of one second.

5. A negative feedback system is given below, where $G_c(s) = K$ and $G(s) = \dfrac{1}{s\left(s + \sqrt{2K}\right)}$

    a. Determine the percent overshoot and settling time for unit step input

    b. For what range of K, the settling time would be ≤ 1 second.

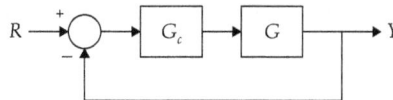

6. A negative feedback system is given below, where $G(s) = 1/[s(s + 2)]$ and $H(s) = (s + 3)/(s + 0.1)$

   a. Determine the steady state error for given step input at $G_p(s) = 1$

   b. Select an appropriate value for $G_p(s)$ so that the steady state error is zero.

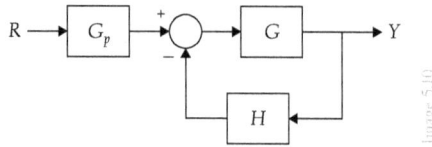

R → G_p → (+ −) → G → Y, with H in feedback

7. Determine the gain setting such that the steady state error is ≤ 0.5 for given input $R(t) = t^2$

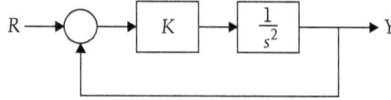

R → K → $\frac{1}{s^2}$ → Y

8. If $G(s) = K/[s(s + p)]$; select a value for "K." and pole "p" such that the percent overshoot is 3.4% and settling time is 2 seconds.

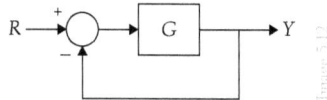

R → (+ −) → G → Y

9. If $G(s) = (s + 12)/(s + 16)$ and $H(s) = (s + 10)/(s + 18)$; determine the gain setting such that the steady state error is zero for given step input.

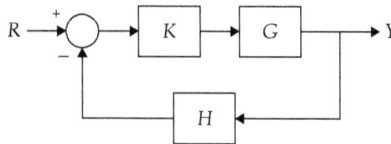

R → (+ −) → K → G → Y, with H in feedback

10. If $G(s) = 1/[s(s + K_1)]$ in the control system given below,

    a. Determine K and K_1 such that the percent overshoot is 6% and settling time is 1.5 seconds

    b. Determine the steady state error due to ramp input

    c. Determine the sensitivity of the transfer function with respect to the parameter K_1

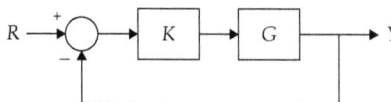

R → (+ −) → K → G → Y

CONTROL SYSTEM STABILITY

6.1. Introduction

A control system's stability is a function of the poles of a transfer function (TF) and their location on an S-plane plot. The poles are obtained by solving the roots of the characteristic equation. Characteristic equation is a polynomial in the denominator of the TF. Finding the roots of characteristic equation becomes a challenge if it is of a higher-order polynomial. The Routh-Hurwitz (R-H) stability criterion provides a method to determine a system's stability without computing the roots of the characteristic equation. An emphasis is given in this chapter to apply this method in the study of a control system's stability.

6.2. Learning Objectives

1. Understand the stability of dynamic systems.

2. Understand the role of poles and their location in finding a system's stability.

3. Apply the R-H method to determine a system's stability.

6.3. Stability of Dynamic Systems

Let the TF of a dynamic system be:

$$T(s) = \frac{1}{s(s+a)(s-a)}$$

The characteristic equation is: $s(s+a)(s-a) = 0$

The roots are: $s = 0, s = -a, s = +a$

These roots are also called as the system's *poles* and can be represented on an S-plane stability plot.

In general, the complex poles are $-a + jy_1$ or $a + jy_2$

If $y_1 = 0 = y_2$, the roots are real. The real poles, $0, -a, a$, are shown in the figure below.

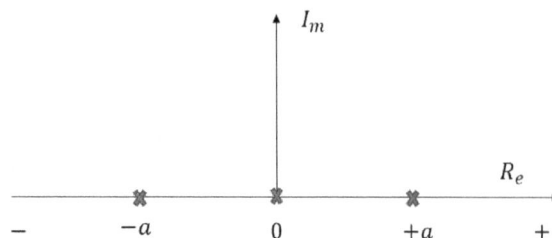

Here, $s = -a$ is stable, and $s = a$ is unstable.

The point, $s = 0$, is neutrally stable.

The response for each case is given below.

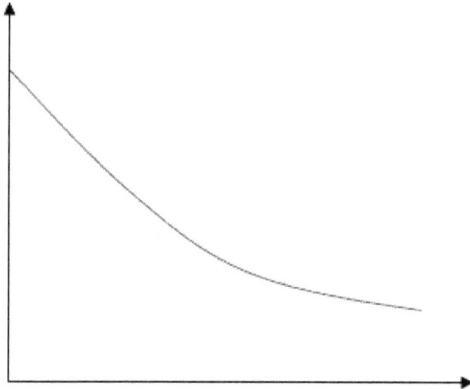

$$s = -a$$
Stable

$$s = 0$$
Neutral

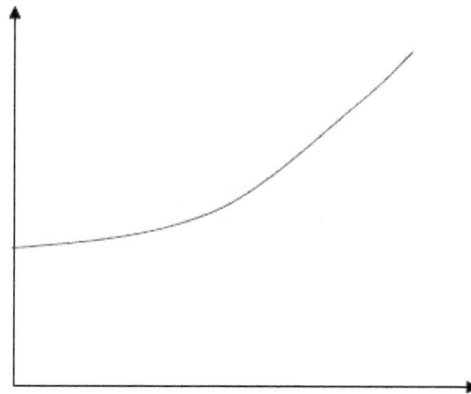

$$s = a$$
Unstable

If $y_1 \neq 0 \neq y_2$, the roots are complex; the respective responses are:

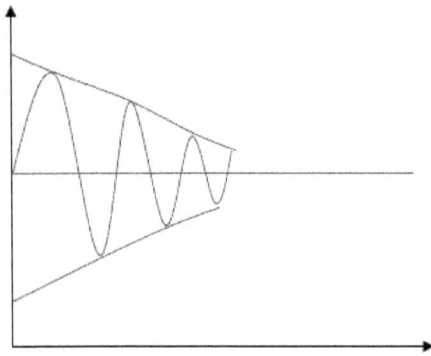

$$s = -a + jy_1$$
Converging Oscillation (Stable)

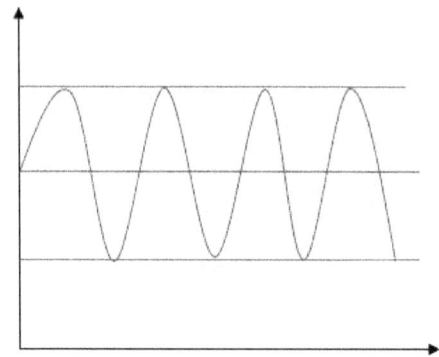

$$s = +jy_1$$
Steady Oscillation
(Marginally Stable)

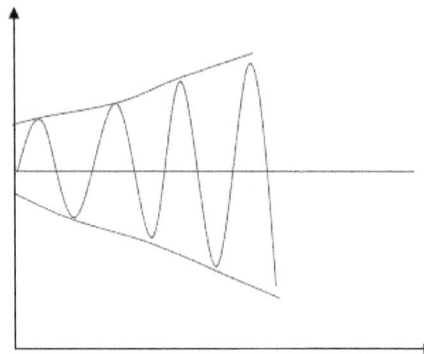

$$s = a + jy_2$$
Diverging Oscillation (Unstable)

Characteristic Equation

Let the system TF, $T(s) = p(s)/q(s)$. The polynomial in the numerator is $p(s)$, and its roots are called the *zeros* of the system. The polynomial in the denominator is $q(s)$, and its roots are called *poles*. The negative poles (real or complex) contribute to the system's stability, whereas the positive poles (real or complex) contribute to the system's instability. The characteristic equation is given as $q(s) = 0$.

$$\text{Let } q(s) = a_n s^n + a_{n-1} s^{n-1} + a_{n-2} s^{n-2} + a_{n-3} s^{n-3} + \cdots + a_1 s + a_0 = 0.$$

The general equation given above represents a nth-order control system. The characteristic equation is used to apply the *R-H* criterion. However, the characteristic equation should be checked first for completeness of nth-order polynomial and no sign change in the coefficients of the polynomial before applying the *R-H* criterion. For example, if the characteristic equation is a fifth-order polynomial, the coefficients of s^5 and all the successive powers of s should be positive and not equal to zero.

6.4. The Routh-Hurwitz Criterion

This criterion states that the number of roots of $q(s)$ with positive real parts is equal to the number of sign changes in the first column of the Routh array. So, a stable system requires all the coefficients in the first column of the Routh array to be positive. An unstable system will have at least one sign change in the first column, indicating at least there is one positive real root. A typical Routh array is given below.

The first row in the array begins with the coefficient of s^n and continues with successive alternate term coefficients. The second row in the array begins with the coefficient of s^{n-1} and continues with successive alternate term coefficients. All other rows in the array need to be computed.

s^n	a_n	a_{n-2}	a_{n-4}	$\cdots$	$\cdots$	a_2	a_0
s^{n-1}	a_{n-1}	a_{n-3}	a_{n-5}	$\cdots$	$\cdots$	a_3	a_1
s^{n-2}							
s^{n-3}							
$\vdots$							
s^2							
s^1							
s^0							

EXAMPLE

$$q(s) = a_3 s^3 + a_2 s^2 + a_1 s + a_0$$

Routh Array

s^3	a_3	a_1	0	
s^2	a_2	a_0	0	
s^1	b_1	b_2		
s^0	c_1	c_2		

To Find b_1

Consider two elements in the same column, just above b_1 and the other two elements in the immediate next column. They are:

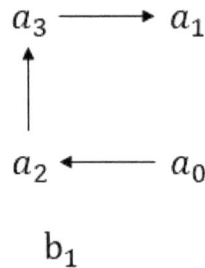

$$a_3 \longrightarrow a_1$$
$$\uparrow$$
$$a_2 \longleftarrow a_0$$
$$b_1$$

$$b_1 = -\left(\frac{a_0}{a_2}\right)a_3 + a_1 = -\frac{1}{a_2}\left[a_3 a_0 - a_2 a_1\right] = -\frac{1}{a_2}\begin{bmatrix} a_3 & a_1 \\ a_2 & a_0 \end{bmatrix}$$

To Find b_2

Skip the elements in the same column, just above b_2. Consider the elements in a column to the left and a column to the right above b_2.

$$b_2 = -\left(\frac{0}{a_2}\right)a_3 + 0 = -\frac{1}{a_2}\left[a_3(0) - a_2(0)\right] = -\frac{1}{a_2}\begin{bmatrix} a_3 & 0 \\ a_2 & 0 \end{bmatrix} = 0$$

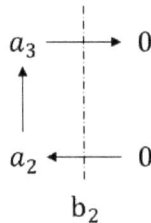

$$a_3 \longrightarrow 0$$
$$\uparrow$$
$$a_2 \longleftarrow 0$$
$$b_2$$

To Find c_1

Similar to b_1, the procedure is to consider four elements in the columns just above c_1.

$$c_1 = -\left(\frac{b_2}{b_1}\right) a_2 + a_0 = a_0 \text{ ; since } b_2 = 0$$

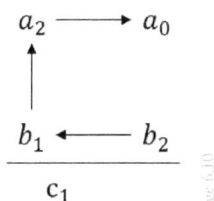

To Find c_2

Similar to b_2, the elements to be considered are:

$$c_2 = -\left(\frac{0}{b_1}\right) a_2 + 0 = 0$$

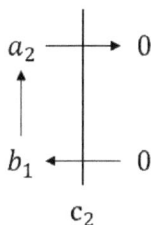

Completed Routh Array

s^3	a_3	a_1	0
s^2	a_2	a_0	0
s^1	b_1	0	
s^0	a_0	0	

$$b_1 = \frac{1}{a_2}(a_1 a_2 - a_3 a_0)$$

For the system to be stable, all the elements in the first column of the Routh array should be positive. In this case, a_3, a_2, b_1, and a_0 should be positive. If a_3 and a_2 are positive, a_0 should be positive. For b_1 to be positive, $a_1 a_2 > a_3 a_0$. It requires a_1 also to be positive because all other coefficients are positive. So, all the coefficients are required to be positive.

In a stability analysis, it is necessary that all the coefficients of a characteristic equation should be positive. If there is any sign change, it is certain that the system is likely to be unstable. The characteristic equation can be inspected for any sign change among coefficients as an initial check. If all the coefficients are positive and not equal to zero, then the R-H method can be applied for further investigation of the system's stability.

The procedure for the R-H method can be classified into four categories. They are: (1) there is no zero element in the first column, (2) there is a zero element in the first column, (3) all elements in a row are zero, and (4) repeated roots—multiple rows with zero elements.

6.4.1. No Zero Element in the First Column

The characteristic equation is:

$$q(s) = s^3 + s^2 + 2s + 24$$

The polynomial is complete. All coefficients exist, and they are positive. Let us apply the R-H criterion.

s^3	1	2	0
s^2	1	24	0
s^1	-22	0	
s^0	24		

In the first column, we see two sign changes, one from s^3 to s^1 and the other from s^1 to s^0. There will be two positive roots, and they will be on the right half of the Splane. So, the system will be unstable.

Roots of the characteristic equation:

$$q(s) = s^3 + s^2 + 2s + 24 = (s+3)(s^2 - 2s + 8)$$

The roots are $s_1 = -3$, $s_{2,3} = 1 \pm j\sqrt{7} = \sigma \pm j\omega$.

Although it satisfies the necessary condition in the initial check (the polynomial is complete, and all coefficients are positive) the Routh array has indicated that the system is unstable.

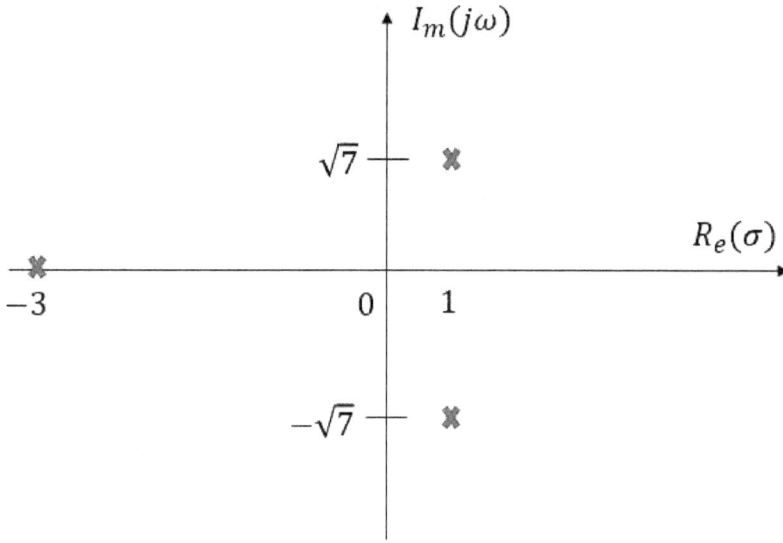

6.4.2. There is a Zero Element in the First Column

If there is a zero in the first column of the Routh array while the other elements in the respective row are nonzero, replace the zero by a small positive number, ϵ, for computing the elements in the first column of the array.

Let the characteristic equation be:

$$q(s) = s^5 + 2 s^4 + 2 s^3 + 4 s^2 + 11 s + 10$$

It satisfies the necessary condition in the initial check that the polynomial is complete and all the coefficients are positive.

Routh Array

s^5	1	2	11	0
s^4	2	4	10	0
s^3	b_1	b_2	0	
s^2	c_1	c_2	0	
s^1	d_1	d_2	0	
s^0	e_1	e_2		

$$b_1 = -(4/2)(1) + 2 = 0$$

$$\text{Let } b_1 = \epsilon$$

$$b_2 = -(10/2)(1) + 11 = 6$$

$$c_1 = -\left(b_2/b_1\right)(2) + 4 = -(6/\epsilon)(2) + 4 = 4(\epsilon - 3)/\epsilon$$

$$\text{Because } \epsilon \ll 3; c_1 = \frac{4(-3)}{\epsilon} = -12/\epsilon$$

$$c_2 = -(0/b_1)(2) + 10 = 10$$

$$d_1 = -(c_2/c_1)b_1 + b_2 = 10\epsilon^2/12 + 6 \approx 6$$

$$d_2 = 0$$

$$e_1 = -(d_2/d_1)c_1 + c_2 = c_2 = 10$$

$$e_2 = 0$$

Completed Routh Array

s^5	1	2	11
s^4	2	4	10
s^3	ϵ	6	
s^2	$-12/\epsilon$	10	
s^1	6	0	
s^0	10	0	

We see that there are two sign changes in the first column. So, the system is unstable with two roots in the right half plane.

6.4.3. ALL Elements in a Row Are Zero

In this case, the elements in a row that includes zero in the first column are also zero. When the roots of characteristic equation are symmetric pairs, such as $(s + \sigma)$ and $(s - \sigma)$ or $(s + j\omega)$ and $(s - j\omega)$, the Routh array has a row with all zero elements. In this type of situation, an auxiliary polynomial is used to determine the stability. The auxiliary polynomial, $U(s)$, is one that results from a row that immediately precedes the row of zeros in the

Routh array. The order of the auxiliary polynomial is always even, indicating the number of symmetrical pairs of roots.

EXAMPLE

Let the characteristic equation be: $q(s) = s^3 + 2s^2 + 4s + 8$.

Routh Array

s^3	1	4	0
s^2	2	8	0
s^1	$\in$	0	
s^0	8	0	

The row s^1 has all zero elements, and the row s^2, just above the row of zeros, is used to form the auxiliary polynomial. Therefore, the auxiliary polynomial is given as:

$$U(s) = 2s^2 + 8 = 0$$

$$= s^2 + 4 = 0$$

$$s^2 = -4; \; s = \pm j2$$

To find the other root of the characteristic equation, divide the polynomial by the auxiliary polynomial.

$$\frac{q(s)}{U(s)} = \frac{s^3 + 2s^2 + 4s + 8}{s^2 + 4}$$

$$
\begin{array}{r}
s + 2 \\
s^2 + 4 \overline{\smash{\big)}\, s^3 + 2s^2 + 4s + 8} \\
\underline{s^3 + 4s} \\
(-) \qquad 2s^2 + 8 \\
\underline{2s^2 + 8} \\
(-) \qquad 0
\end{array}
$$

Therefore, $q(s) = (s + 2)(s + j2)(s - j2)$.

The roots (or poles) of $q(s)$ are $-2, j2, -j2$.

S-PLANE PLOT

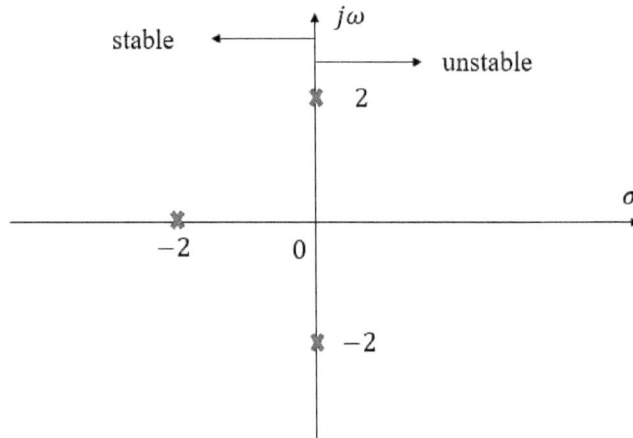

Because the poles, $\pm j2$, are on the vertical axis that divides the stable and unstable regions, the system is neither stable nor unstable. It is classified as marginally stable. For all practical purposes, it is considered unstable.

6.4.4. Repeated Roots—Multiple Rows With Zero Elements

In the previous case, we had one row of zeros on the Routh array, resulting in one auxiliary polynomial with one pair of symmetric roots on the vertical axis of the s-plane plot. In this case, we will see multiple rows of zeros on the Routh array, resulting in more than one auxiliary polynomial with several repeated roots on the vertical axis of the s-plane plot.

EXAMPLE

Let the characteristic equation be:

$$q(s) = s^5 + s^4 + 2s^3 + 2s^2 + s + 1$$

Routh Array

s^5	1	2	1	0
s^4	1	2	1	0
s^3	$\in$	$\in$	0	
s^2	1	1	0	
s^1	$\in$	0		
s^0	1			

Here, the rows s^3 and s^1 contain all zero elements. The value $\in$ is used at appropriate places in order to complete the following rows on the Routh array.

The auxiliary polynomial for the row s^1 is:

$$s^2 + 1 = 0; \; s^2 = -1; \; s = \pm j; \; s_1 = +j; \; s_2 = -j$$

The auxiliary polynomial for the row s^3 is:

$$s^4 + 2s^2 + 1 = 0 = (s^2 + 1)^2$$

$$(s^2 + 1)(s^2 + 1) = 0$$

$$(s^2 + 1) = 0; \; s_3 = +j; \; s_4 = -j$$

$$(s^2 + 1) = 0; \; s_5 = +j; \; s_6 = -j$$

There are three pairs of repeated symmetric roots on the vertical axis, leading to marginal instability. The system is unstable.

EXAMPLE

Given that $q(s) = s^5 + s^4 + 15\,s^3 + 10\,s^2 + 44\,s + 24$:

Find all the roots and check for stability.

Routh Array

s^5	1	15	44
s^4	1	10	24
s^3	5	20	0
s^2	6	24	0
s^1	$\in$	0	
s^0	24		

The auxiliary polynomial is: $6\,s^2 + 24 = 0$.

$$U(s) = s^2 + 4 = 0; \; s = \pm j2$$

There are two roots on the vertical axis, and hence the system is marginally stable and considered unstable for engineering applications.

To explore for other possible unstable roots, divide $q(s)$ by the auxiliary polynomial and reduce the order of the polynomial to cubic:

$$\frac{q(s)}{U(s)} = \frac{s^5 + s^4 + 15\, s^3 + 10\, s^2 + 44\, s + 24}{s^2 + 4}$$

$$s^3 + s^2 + 11\, s + 6$$

$$s^2 + 4 \overline{\big)\; s^5 + s^4 + 15\, s^3 + 10\, s^2 + 44\, s + 24}$$

$$\underline{s^5 + \qquad 4\, s^3}$$

$$s^4 + 11\, s^3 + 10\, s^2$$

$$\underline{s^4 + \qquad\quad + 4\, s^2}$$

$$11\, s^3 + 6\, s^2 + 44\, s$$

$$\underline{11\, s^3 + \qquad 44\, s}$$

$$6\, s^2 + \qquad 24$$

$$\underline{6\, s^2 + \qquad 24}$$

$$0$$

The reduced polynomial is: $s^3 + s^2 + 11\, s + 6 = 0$.

Applying the Routh criterion for this polynomial:

Routh Array

s^3	1	11
s^2	1	6
s^1	5	0
s^0	6	

There are no sign changes in the first column, and hence there are no further unstable roots.

The roots of cubic polynomial can be easily solved, and they are: -0.56, $-0.22 \pm j6.54$. So, the characteristic equation has five roots with one negative real root and two complex roots with a negative real part and two imaginary roots. The imaginary roots with no real part cause the system to be marginally stable.

EXAMPLE

For a given control system, find the range of k and a for which the system is stable.

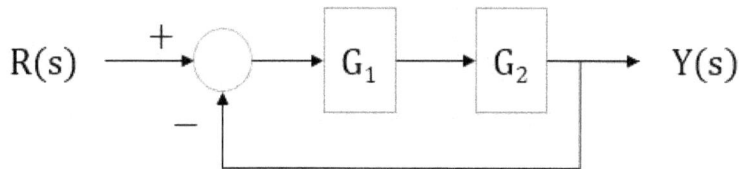

$$G_1(s) = \frac{k(s+a)}{s+1}$$

$$G_2(s) = \frac{1}{s(s+2)(s+3)}$$

$$T(s) = \frac{G_1 G_2}{1 + G_1 G_2} = \frac{k(s+a)}{\left[s(s+1)(s+2)(s+3) + k(s+a) \right]}$$

The characteristic equation is given by the denominator.

$$q(s) = s(s+1)(s+2)(s+3) + k(s+a) = s^4 + 6s^3 + 11\, s^2 + (k+6)s + ka$$

Routh Array

s^4	1	11	ka
s^3	6	$(k+6)$	0
s^2	b_1	b_2	0
s^1	c_1	0	0
s^0	d_1	0	

$$b_1 = -\frac{(k+6)}{6} + 11 = \frac{60 - k}{6}$$

$$b_2 = ka$$

$$c_1 = -\frac{b_2}{b_1}(6) + (k+6) = \frac{1}{(60-k)}\left[(k+6)(60-k) - 36\,ka\right]$$

$$d_1 = 0 + b_2 = ka$$

For the system to be stable, the elements b_1, c_1, and d_1 in the first column should be positive.

For b_1 to be positive, $k < 60$.

For d_1 to be positive, $k > 0$ and $a > 0$.

The range of a depends on the value of k chosen in the range of $0 < k < 60$.

For $c_1 > 0$, $(k+6)(60-k) - 36ka > 0$ OR $(36\,ka) < (k+6)(60-k)$.

So, $a < (k+6)(60-k)/(36\,k)$.

For $k = 1$, $a < 11.47$

$$k = 59,\ a < 0.03$$

The range of $k = 1 \le k \le 59$

The range of $a = \begin{cases} k = 1;\ 0 \le a \le 11.47 \\ k = 59;\ 0 \le a \le 0.03 \end{cases}$

6.5. Summary

The R-H criterion is useful in determining whether a control system is stable or unstable without finding the roots of the characteristic equation, especially if it is a higher-order polynomial. The system is stable if the poles are negative, real or complex. The system is unstable if the poles are positive, real or complex. If a pole is real and zero, placed at the origin, the system is neutrally stable. If a pole is only a pair of imaginary numbers on the vertical axis at the origin, the system is marginally stable; however, it is considered unstable for practical applications. For a marginally stable system, all the elements in one row of the Routh array will be zero. The auxiliary polynomial is used in such a case to determine the system's stability.

6.6. Assessment

1. The characteristic equation is:

   a. The number of the controller's TF

   b. The denominator of the controller's TF

    c. The numerator of the total (system)'s TF

    d. The denominator of the total (system)'s TF

2. The roots of characteristic equation are called:

    a. Zeros

    b. Poles

    c. Complex zeros

    d. None of the above

3. For stability, the roots of characteristic equation should be on:

    a. The left half plane of the S-plane plot

    b. The right half plane of the S-plane plot

    c. The real axis of the S-plane plot

    d. The vertical axis of the S-plane plot

4. For stability, the coefficients of s in the characteristic equation should be:

    a. All positive

    b. All negative

    c. At least one positive and one negative

    d. An equal number of positive and negative

5. For stability, the Routh array should be as follows:

    a. All values in the first column be positive

    b. All values in the second column be positive

    c. At least one negative value in the first column

    d. None of the above

6. Regarding the first two rows on the Routh array:

    a. One row has odd powers, and the other has even powers

    b. The coefficients of s are used to complete these two rows

    c. The first row begins with the highest power of the characteristic equation

    d. All of the above

7. On the given Routh array, the value b_1 is:

s^3	1	8	0
s^2	2	4	0
s^1	b_1	b_2	
s^0	c_1		

a. −12
b. 6
c. 0
d. 3

8. The value of c_1 on the Routh array is:

a. 4
b. 0
c. 8
d. −16

9. An auxiliary polynomial is used for stability if:

a. One of the elements on the first column of the Routh array is zero
b. One of the columns on the Routh array is zero
c. One of the rows on the Routh array has all zeros
d. One of the elements on the first row of the Routh array is zero

10. The auxiliary polynomial:

a. Is the row preceding the row of zeros
b. Is an even-order polynomial
c. Has symmetric roots
d. All of the above

6.7. Practice Problems

1. Answer all the assessment questions in Section 6.6

2. Determine the stability of the systems described by the following characteristic equations

   a. $s^2 + 5s + 2 = 0$

   b. $s^3 + 2s^2 - 6s + 20 = 0$

   c. $s^5 + s^4 + 2s^3 + s + 6 = 0$

3. A control system has the characteristic equation of $s^3 + 15s^2 + 2s + 40 = 0$ Determine the stability of the system by Routh-Hurwitz method.

4. A control system is shown below. Determine the range of value to the gain "K" for system stability.

$$G(s) = \frac{K(s + 10)}{[s(s + 3)(s^2 + 4s + 8)]}$$

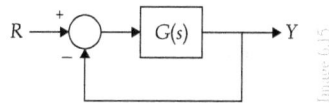

5. Determine the range of value for "K", for system stability.

$$G(s) = \frac{1}{s(s + 1)}; \quad H(s) = \frac{1}{(s + 4)}$$

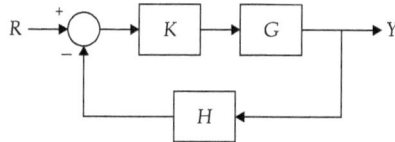

6. Determine the stability of the system with a transfer function as given below.

$$T(s) = \frac{(s + 3)}{(s^3 + 4s^2 + 8s + 4)}$$

7. Determine the range of values for "K", for system stability

$$G(s) = \frac{1}{(s^3 + s^2 + 3s + 2)}$$

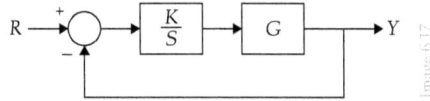

8. Determine the stability of the control system given below.

$$G_1(s) = \frac{1}{(s+1)}; \quad G_2(s) = \frac{1}{(s^3 + 4s^2 - 2s + 10)}; \quad H(s) = \frac{3}{s}$$

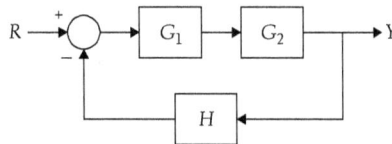

9. Determine the stability of the system given below.

$$G_1(s) = \frac{1}{(s+1)^2}; \quad G_2(s) = \frac{1}{(s^2 + 4)}$$

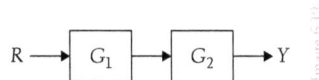

10. Determine the range of values for "K", for system stability

$$G_1(s) = \frac{K}{(s+10)}; \quad G_2(s) = \frac{1}{(s-1)(s+2)}$$

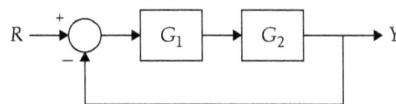

ROOT LOCUS
METHOD

7.1. Introduction

The transfer function (TF) of a control system is generally a function of the controller gain, K. The poles and zeros obtained from the TF are for a given value of K and will change when the controller gain changes. The loci of the poles and how they travel with respect with the variation of K from zero to infinity can be shown on a diagram, called the *root locus diagram*. The process and various steps to draw the diagram are discussed in this chapter. A study on a commonly used industrial control based on proportional, integral, and derivative (PID) controllers is also provided in this chapter.

7.2. Learning Objectives

1. Understand the roots (poles and zeros) of a TF and how they travel on the S-plane plot with respect to changes in the controller gain, K.

2. Learn to sketch the root locus diagram that shows the path or loci from a pole to a zero at a finite distance or at infinity.

3. Apply the three term PID parameters in a controller design to achieve the desired performance.

7.3. Loci of the Roots

The root locus method was developed by Walter R. Evans, who received his engineering education in electrical engineering at Washington University in St. Louis in 1941. He invented the root locus method in 1948 before he completed his master's degree in electrical engineering at the University of California, Los Angeles, in 1951.

Closed Loop

Loci is a path traveled by a pole of the TF and is called the *root locus diagram* on an S-plane plot. In a given closed-loop control system, K is a controller parameter, and $P(s)$ is the plant's TF.

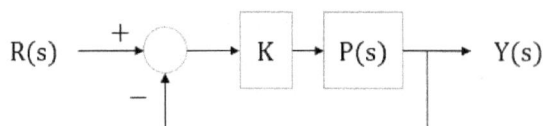

The TF: $T(S) = \dfrac{KP}{1 + KP}$

The characteristic equation is $q(s) = 1 + KP = 0$, and its roots are the poles.

If $P(s) = \dfrac{N(s)}{D(s)}$; $T(S) = \dfrac{KN(s)}{D(s) + KN(s)}$,

$$q(s) = D(s) + KN(s) = 0$$

for $K = 0$, $q(s)$ reduces to $D(s) = 0$, and its roots are the poles of the closed-loop system.

Open Loop

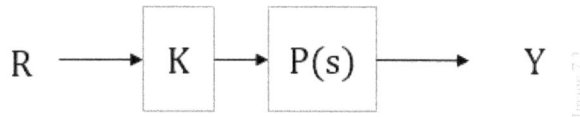

$$T(s) = KP(s); \; P(s) = \dfrac{N(s)}{D(s)}$$

The characteristic equation is $D(s) = 0$, and its roots are the poles of an open-loop system. These poles are equal to the poles of the closed-loop system at K and equal to zero.

Open-loop zeros are the roots of $N(s) = 0$.

In a closed-loop system, if K is very large and tends to infinity, $D(s) + KN(s) \approx KN(s) = 0$. So, $N(s) = 0$, and its roots are the poles of the closed-loop system. Compared with an open-loop system, the poles of the closed-loop system with $K \to \infty$ are equal to the zeros of the open-loop system.

So, at $K = 0$, the closed-loop poles = open-loop poles.

At $K = \infty$, the closed-loop poles = open-loop zeros.

In other words, when the control parameter K varies from zero to infinity, the locus of a closed-loop pole starts from an open-loop pole and ends at an open-loop zero. So, the journey of a closed-loop pole can be studied when the open-loop poles and zeros are known. The journey from an open-loop pole to an open-loop zero is defined as the root locus or journey of roots. If the path taken by the roots in this journey remains completely in the left half plane, the system is said to be stable for all values of K, and the system is not sensitive to the variation of the control gain K. However, if some of the path is on the right half plane, the system will be unstable, and we can determine the range of K for instability.

Also, K should be positive for the path or loci to exist for travel. If K is negative, no path or loci exists for the roots to travel.

EXAMPLE

Given that $T(s) = \dfrac{KP(s)}{KP(s)+1}$; $P(s) = \dfrac{s+1}{s(s+2)}$,

$$T(s) = \frac{K(s+1)}{s(s+2)+K(s+1)}$$

The characteristic equation: $q(s) = s(s+2) + K(s+1) = 0$

$$s^2 + (K+2)s + K = 0$$

The poles: $s_{1,2} = (1/2)\left[-(K+2) \pm \sqrt{(K+2)^2 - 4K}\right]$

The zeros: $K(s+1) = 0 \rightarrow (s+1) = 0 \rightarrow s = -1$

Because $K = 0$ at the poles, we can find the poles by setting $K = 0$ as:

$$s_{1,2} = (1/2)\left[-2 \pm \sqrt{4-0}\right] = -1 \pm 1 = 0, -2$$

Now, let us plot the poles and zeros on the S-plane plot.

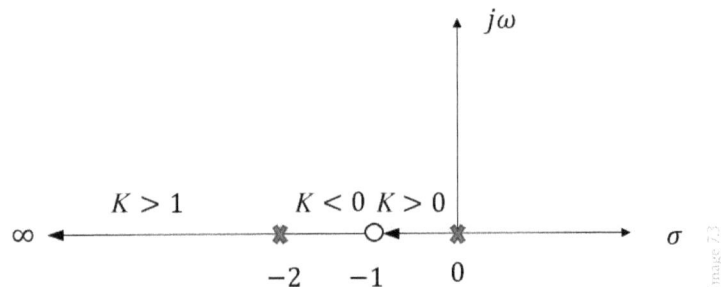

In this plot, the zero at $s = -1$ is represented by a small circle on the real axis, while the poles at $s = 0$ and $s = -2$ are represented by an x on the real axis. The first path of travel starts from the pole at origin and ends at zero at $s = -1$. The second path of travel starts from the pole at $s = -2$ and travels to the zero at infinity. For the second path, a zero is assumed to be at infinity because there is no known zero in its path. In the region $s = -1$ to $s = -2$, there is no path, and K is negative.

For example, let $K = 1.5$, then $s_{1,2} = -0.5, -3$. The first root is on the first path of travel, while the second root is on the second path of travel. Let $K = -1.5$, and then the roots are $s_{1,2} = -1.5, 1$. The first root is in the region where there is no path, and the other root is on the positive half plane, indicating instability. This example highlights that K should be positive for stability, and it is the required condition for the journey of roots starting from a pole ($K = 0$) and ending at a zero ($K = \infty$).

In this example, the poles and zeros are real values. Generally, they can be complex values. Also, it will be challenging to find all the poles and zeros for each value of K. To predict the journey of roots without finding them, one can use a set of rules that were provided by Walter R. Evans, the inventor of this method.

7.4. Process for a Root Locus Plot

For the real values of poles and zeros, the root locus diagram requires the knowledge of number of loci (or paths) for traveling on a real axis. For the complex value of poles and zeros, more information is required, and they are the asymptotes, crossover point, breakaway point, and angle of departure. The process to determine each step in this journey of the roots is given below.

7.4.1. Step 1: Zeros, Poles, and Loci
Given the plant's TF $P(s)$, the closed-loop system TF is:

$$T(s) = \frac{KP(s)}{KP(s)+1} \text{ Let } P(s) = \frac{N(s)}{D(s)}$$

$N(s)$ = numerator polynomial in s, and its roots are the zeros.

$D(s)$ = denominator polynomial in s, and its roots are the poles.

Let number of zeros = m and number of poles = n.

Then, number of loci = n; the number of poles. Number of loci traveling to infinity = $n - m$.

Each loci travels from a pole to a zero located at a finite distance. If there is no zero in its path, a zero is assumed at infinity, and the loci will travel to infinity.

7.4.2. Step 2: Real Axis Loci

A locus exists in a region (between two poles or between a pole and a zero) if the number of poles and zeros to the right of an arbitrary point in the region is an odd number. The controller gain, K, is positive in this region (K > 0).

A locus does not exist in a region if the number of poles and zeros to the right of an arbitrary point is an even number. The controller gain is negative in this region (K < 0).

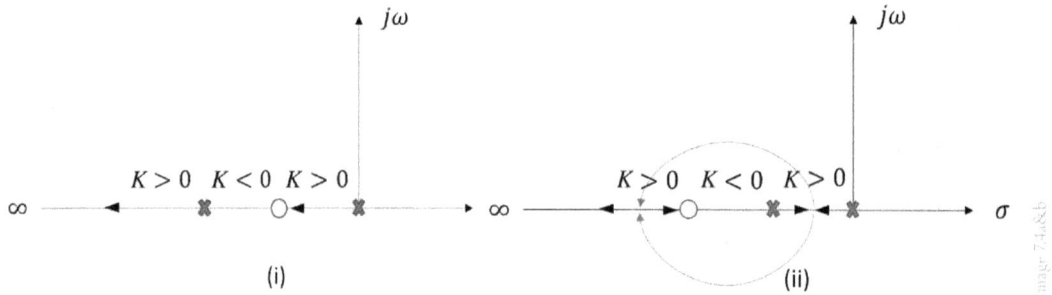

(i)   (ii)

In Figure 1, the loci's travel is simple. In Figure 2, the loci between the two poles meet at a point and break away toward the zero. Because no path exists in the region between the pole and zero, the loci takes a detour of going over that region and enters or breaks in on the real axis, with one branch traveling to zero in the vicinity and the other branch traveling to zero at infinity.

7.4.3. Step 3: Loci Travel to Infinity

If a zero at infinity is not on a real axis but is located at some point on an S-plane, the loci needs guidance to travel. The required parameters for guidance are:

   a. Asymptotes

   b. Crossover points on the imaginary axis for both loci and asymptotes

   c. Breakaway or break-in point on the real axis

(A) ASYMPTOTES

Center of asymptote: $\sigma_A = \dfrac{\Sigma(\text{poles}) - \Sigma(\text{zeros})}{n - m}$

Angles of asymptotes: $(\phi_A) = \left(\dfrac{2k + 1}{n - m}\right)(180)$

$$0 \leq k \leq (n - m - 1)$$

(B) CROSSOVER POINTS

Crossover points are of two types: asymptote crossover points and loci crossover points. The crossover points are defined as points where the asymptote and loci cross the diagram's vertical axis.

$$\text{Asymptote crossover point} = \pm j \left| \sigma_A \right| \tan(\phi_1); (\phi_1) = (\phi_A) \text{ at } k = 0$$

The loci crossover point on the vertical axis represents the condition of marginal stability. So, these points need to be determined by the Routh-Hurwitz criterion.

In the Routh array, find K so that one of the rows has all zeros. Then, find the roots of the auxiliary polynomial, which are the loci's crossover points.

(C) BREAKAWAY OR BREAK-IN POINTS

The breakaway point is the point on a real axis where two loci meet and depart (break away) from the real axis. The break-in point is the point on a real axis where two loci from the S-plane arrive (break in) to the real axis. $K = K_{max}$ at this point.

The characteristic equation is: $1 + KP(s) = 0$.

$$K = -\left(\frac{1}{P(s)} \right)$$

For $K = K_{max}$, $\dfrac{dK}{ds} = 0$

Solving for roots of the resulting polynomial, the root in the appropriate region of loci gives the breakaway or break-in points.

7.4.4. Step 4: Complex Poles and Zeros

When a locus departs from a complex pole, the angle of departure needs to be determined. Similarly, when a locus arrives at a complex zero, the angle of arrival must be determined.

1. **Angle of departure (θ_d)**

   At a complex pole: $s = -a + jb$

$$P'(s) = \left[(s + a - jb)P(s) \right]_{s=(-a+jb)} = \frac{N'(s)}{D'(s)}$$

$$\angle P'(s) = \angle N' - \angle D'$$

$$\angle s = -\tan^{-1}\left(\frac{b}{a}\right) = \begin{cases} 0 \text{ for } b = 0 \\ 90° \text{ for } a = 0 \end{cases}$$

Angle of departure: $\theta_d = \pm\left[\angle P'(s) + 360\right]$ if $\angle P'(s)$ is negative

$\theta_d = \pm\left[\angle P'(s) + 180\right]$ if $\angle P'(s)$ is positive

2. **Angle of arrival (θ_a)**

At a complex zero: $s = -c + jd$

$$P''(s) = \left[P(s)/(s + c - jd)\right]_{s=(-c+jd)} = N''/D''$$

The angle:

$$\angle P''(s) = \angle N'' - \angle D''$$

Angle of arrival: $\theta_a = \pm\left[\angle P''(s) + 360\right]$

NOTE: The poles, zeros, asymptotes, and departure angle are functions of the open-loop TF, $P(s)$.

The crossover points for the loci, breakaway point, and break-in point are functions of the closed-loop TF, $T(s)$.

7.5. Examples

Example 1

Given that $P(s) = \dfrac{1}{s(s + 2)(s + 4)}$, draw the root locus diagram

1. OPEN-LOOP TF, $P(s) = 1/\left[s(s + 2)(s + 4)\right]$

Number of zeros, $m = 0$

Number of poles, $n = 3$; $s = 0, -2, -4$

Number of loci $= n = 3$

Number of loci traveling to infinity $= n - m = 3$

2. EXISTENCE OF LOCI ON A REAL AXIS

s_1 to s_2: sum of poles and zeros to the right of an arbitrary point chosen in this region is odd (one pole).

So, the path exists in this region, and K is positive.

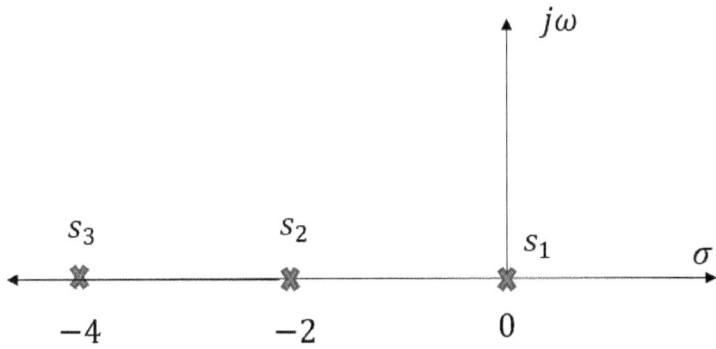

s_2 to s_3: sum of poles and zeros to the right of an arbitrary point chosen in this region is even (two poles). So, no path exists in this region, and K is negative.

s_3 to ∞: sum of poles and zeros to the right of an arbitrary point chosen in this region is odd (three poles). So, the path exists in this region, and K is positive.

3. LOCI TRAVEL

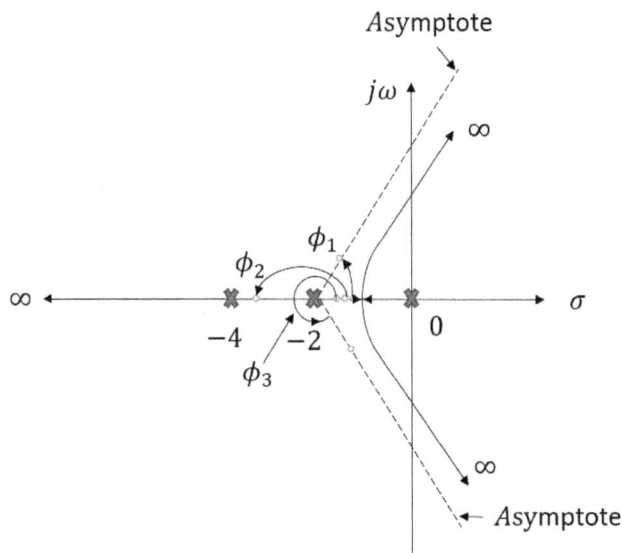

The loci from $s = 0$ and $s = -2$ come together and break away to a zero at infinity. To guide the path of this travel, asymptotes are used. The locus follows these asymptotes in its journey to a zero at infinity.

$$\text{Center of asymptotes: } \sigma_A = \frac{\Sigma P - \Sigma Z}{n - m} = \frac{(0 - 2 - 4) - (0)}{3} = -2$$

$$\text{Angle of asymptotes: } \phi_A = \left|\frac{2k + 1}{n - m}\right|(180); \ 0 \le k \le (n - m - 1)$$

$$\phi_A = \left|\frac{2k + 1}{3}\right|(180) = (2k + 1)(60); \ 0 \le k \le 2$$

$$k = 0; \ \phi_1 = 60°$$

$$k = 1; \ \phi_2 = 180°$$

$$k = 2; \ \phi_3 = 300°$$

Crossover Point

The characteristic equation of a closed-loop TF: $1 + KP(s) = 0$

$$s(s + 2)(s + 4) + K = 0$$

$$s^3 + 6s^2 + 8s + K = 0$$

s^3	1	8	0
s^2	6	K	0
s^1	b_1	0	
s^0	c_1		

$$b_1 = -\frac{K}{6} + 8 = (1/6)(48 - K) \qquad\qquad c_1 = K$$

For stability, b_1 and c_1 should be positive. So, K should be positive but less than 48:

$0 < K < 48$. If $b_1 = 0$, the row S will have all zero; the system will be marginally stable.

For $K = 48$, $b_1 = 0$:

The auxiliary polynomial is $6s^2 + K = 0$; $6s^2 + 48 = 0$.

$$S = \sqrt{-\frac{48}{6}} = \pm j2.828;$$ these are the loci crossover points.

The asymptote crossover points are $\pm j|\sigma_A|\tan\phi_1 = \pm j2\tan60 = \pm j3.46$.

Breakaway Point

$$1 + KP(s) = 0; \quad K = -1/P(s) = -\left(s^3 + 6s^2 + 8s\right) = 0$$

$$dK/ds = 0; \quad 3s^2 + 12s + 8 = 0$$

$$s = (1/6)\left[-12 \pm \sqrt{144 - 96}\right] = -0.845, -3.155$$

Because -3.155 lies on the region where there are no loci, $s = -0.845$ is the viable breakaway point. So, the loci from $s = 0$ and $s = -2$ travel toward each other, break away at $s = -0.845$, and follow the asymptotes to the zero at infinity. In this process, the loci cross the vertical axis at $s = \pm j2.828$, and the asymptote crosses the vertical axis at $\pm j3.46$.

Because the pole is not a complex pole, it is not required to find the angle of departure.

Example 2

Given that $P(s) = (s + 2)/s(s + 1)$, draw the root locus diagram.

Number of zeros, $m = 1$; $s = -2$

Number of poles, $n = 2$; $s = 0, -1$

Number of loci $= n = 2$

Number of loci travel to infinity $= n - m = 1$

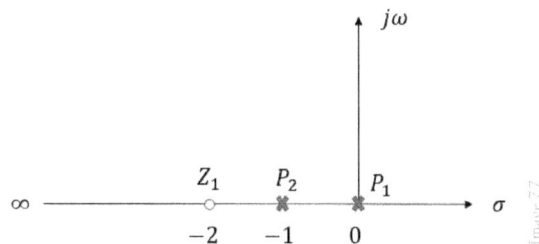

Real Axis Loci

There is a path between P_1 and P_2. There is no path between P_2 and Z_1. There is a path between Z_1 and infinity. The loci from P_1 and P_2 travel toward each other and break away. Then, they take a detour over the zero and break in on the real axis in the region Z_1 to ∞. One branch goes to zero at $s = -2$, and the other branch goes to zero at infinity, as shown in the diagram.

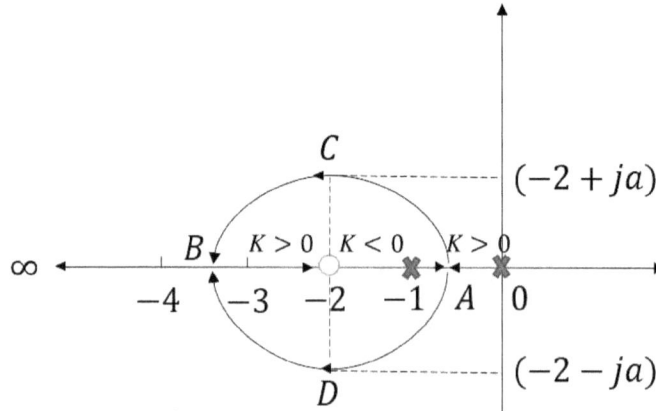

Breakaway and Break-In Points

$$K = -1/P(s) = -s(s+1)/(s+2) = -(s^2+s)/(s+2)$$

$$dK/ds = 0 = -\left[\frac{(s+2)(2s+1) - (s^2+s)(1)}{(s+2)^2}\right]$$

$$s^2 + 4s + 2 = 0$$

$$s = \left(-2 \pm \sqrt{2}\right) = -0.5858, \ -3.4142$$

$s = -0.585$ is the breakaway point, while $s = -3.415$ is the break-in point.

Let the maximum height of the detour path be $-2 \pm ja$.

The characteristic equation is:

$$1 + KP(s) = 0$$

$$s(s+1) + K(s+2) = 0; \quad s^2 + (K+1)s + 2K = 0 \rightarrow (1)$$

For the roots, $s = -2 \pm ja$. The desired characteristic equation is:

$$(s + 2 - aj)(s + 2 + aj) = 0$$

$$s^2 + 4s + (4 + a^2) = 0 \rightarrow (2)$$

Comparing equations (1) and (2):

$$K + 1 = 4; K = 3$$

$$4 + a^2 = 2K = 6$$

$$\text{So, } a^2 = 2; a = \sqrt{2}$$

The coordinates of loci at $s = -2$ are, $s = -2 \pm j\sqrt{2}$.

On the root locus diagram, points A and B are at an equal distance of 1.4142 (or $\sqrt{2}$) from the zero at $s = -2$. Points C and D are at an equal distance of $\sqrt{2}$ from the pole at $s = -2$.

So, the shape of the root locus is a circle.

K-Value at Breakaway and Break-In Points

Because $K = -(s^2 + s)/(s + 2)$

At the breakaway point, $s = -0.585; K = 0.17$

At the break-in point, $s = -3.415; K = 5.83$

Example 3

Given that $P(s) = \dfrac{1}{\left[s(s + 4)(s^2 + 8s + 32) \right]}$

Draw the root locus diagram.

Zeros, Poles, and Loci

Number of zeros, $m = 0$

Number of poles, $n = 4$; $s_1 = 0$, $s_2 = -4$; $s_{3,4} = 1/2\left[-8 \pm \sqrt{64 - 128}\right] = -4 \pm j4$

Number of loci $= n = 4$

Number of loci travel to infinity $= n - m = 4$

Real axis loci:

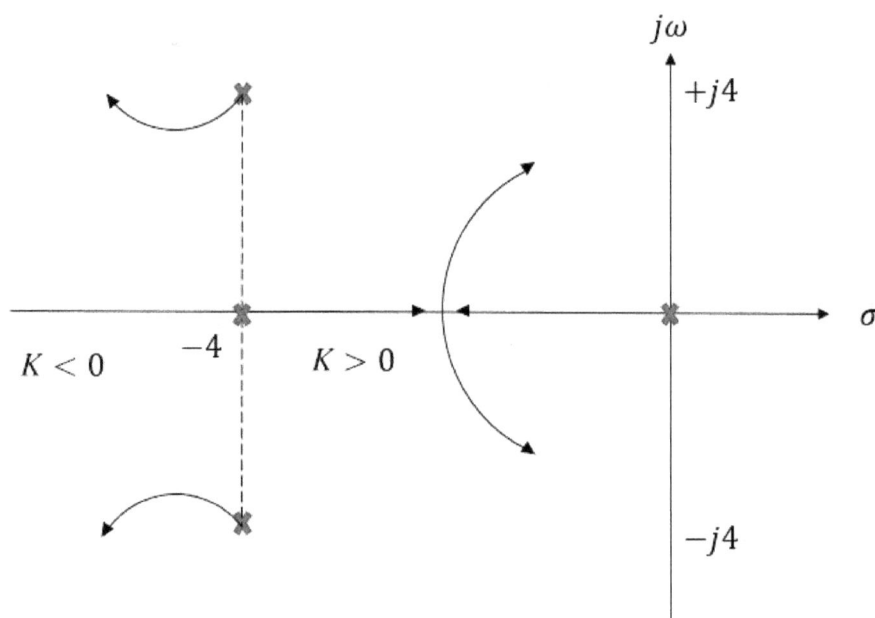

There is a locus between $s = 0$ and $s = -4$. There are no loci in the region $s = -4$ and $s = \infty$. The loci between the poles $s = 0$ and $s = -4$ travel toward each other and break away to a zero at infinity. The loci from complex poles begin their journey with a departure angle and travel to a zero at infinity. Because the path to infinity is not along the real axis, they need the guidance of asymptotes.

Loci Travel

Asymptotes:

Center of asymptotes: $\sigma_A = \dfrac{\sum P - \sum Z}{n-m} = (1/4)\left[0 - 4 - 4 + j4 - 4 - j4\right] = -3$

Angle of asymptotes: $\phi_A = \left(\dfrac{2k+1}{n-m}\right)(180°) = \left(\dfrac{2k+1}{4}\right)(180°) = (2k+1)(45°)$

$$0 \leq k \leq (n-m-1); \; 0 \leq k \leq 3$$

k	0	1	2	3
ϕ_A	45°	135°	225°	315°

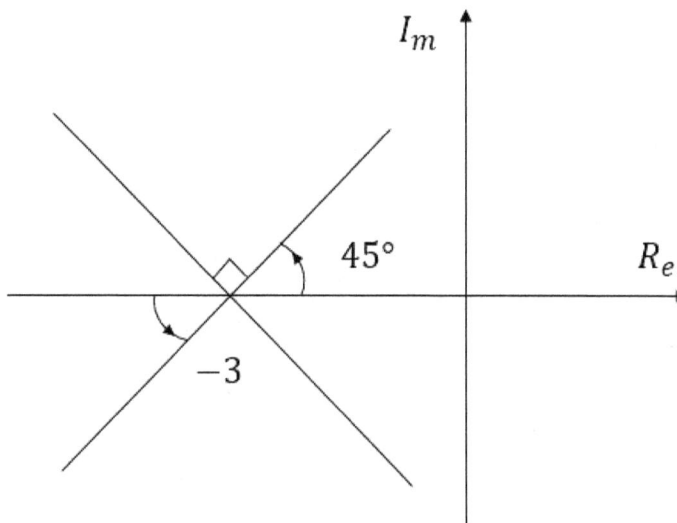

Crossover Point

The characteristic equation is given by:

$$1 + KP(s) = 0$$

$$s(s+4)(s^2 + 8s + 32) + K = 0; \; s^4 + 12s^3 + 64s^2 + 128s + K = 0$$

Routh Array

s^4	1	64	K
s^3	12	128	0
s^2	b_1	b_2	
s^1	c_1	c_2	
s^0	d_1		

$$b_1 = -(128/12)(1) + 64 = 160/3$$

$$b_2 = K$$

$$c_1 = 128 - (9K/40)$$

$$c_2 = 0$$

$$d_1 = K$$

For stability, b_1, c_1, and d_1 should be positive.

So, $K > 0$ and $\dfrac{9K}{40} < 128$; or $K < 568.89$.

For $K = 568.89$, $c_1 = 0$, and hence the row of s^1 has all zeros.

Then, the auxiliary polynomial is given by $b_1 s^2 + b_2 = 0$:

$$s = \sqrt{-\left(b_2/b_1\right)} = \sqrt{-3K/160} = \pm j(3.27)$$

These are the crossover points of loci.

The crossover point of asymptotes $= \pm\,|\,\sigma_A\,|\tan\phi_1 = \pm 3\tan45 = \pm 3$.

Breakaway Point
The characteristic equation is: $1 + KP(s) = 0$:

$$K = -1/P(s) = -(s^4 + 12s^3 + 64s^2 + 128s)$$

The value of s at $K = K_{max}$ gives the breakaway point.

So, $\dfrac{dK}{ds} = 0$; $4s^3 + 36s^2 + 128s + 128 = 0$.

Let $F(s) = dK/ds$; because $F(s)$ is cubic, we can solve for a root by the graphical method.

s	0	-1	-2
$F(s)$	128	32	-16

The breakaway point is: $\sigma_b = -1.58$.

Alternately, one can find the roots of this cubic polynomial through MATLAB and choose the real value of the root that is within the region of the loci.

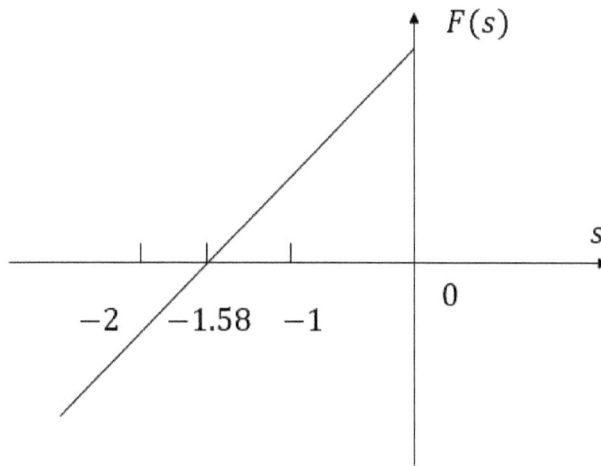

Departure Angle

Let us find the departure angle at the complex pole: $-4 + j4$

$$P(s) = \dfrac{1}{\left[s(s+4)(s+4+j4)(s+4-j4)\right]}$$

At $s = -4 + j4$:

$$P'(s) = \left[(s+4-j4)P(s)\right]\bigg|_{s=-4+j4}$$

$$= \left[\frac{1}{s(s+4)(s+4+j4)} \right]_{s=-4+j4}$$

$$= \frac{1}{\left[(-4+j4)(j4)(j8) \right]}$$

$$P'(s) = N'/D';$$

$$\angle P' = \angle N' - \angle D'$$

$$= (0) - (-45 + 90 + 90) = -135°$$

$$\theta_d = \pm(-135 + 360) = \pm225°$$

For the pole, $-4+j4$; $\theta_d = 225°$, and for the pole, $-4-j4$; $\theta_d = -225°$.

Root Locus Diagram

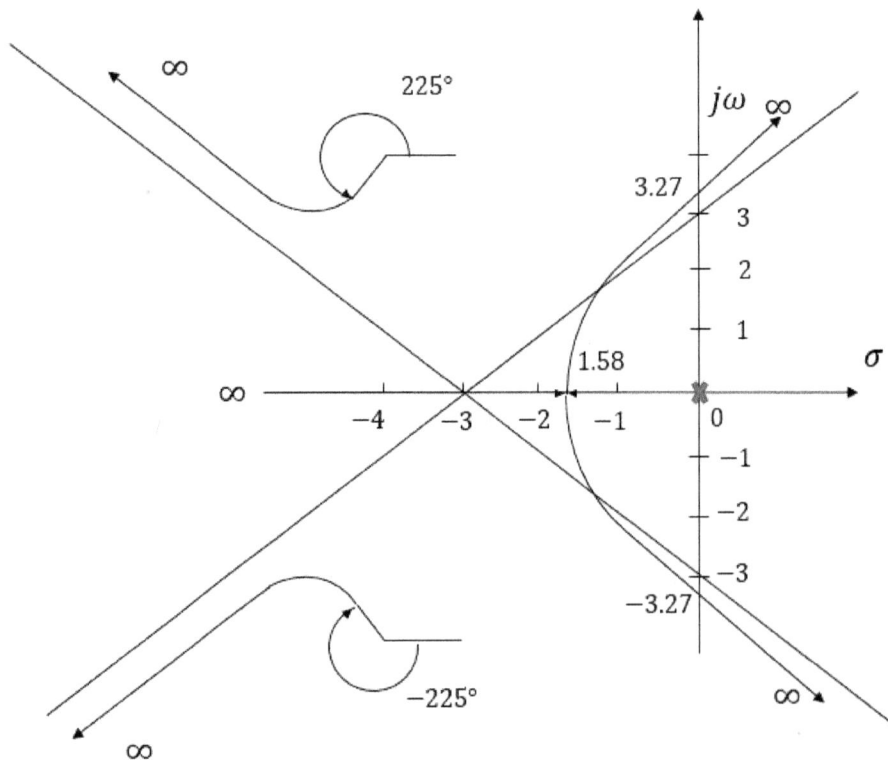

7.6. PID Controllers

"PID" is a three terms controller that is commonly used in industrial process. The three terms are (1) proportional, (2) integral, and (3) derivative.

$$G_c(s) = K_P + \left(K_I \middle/ s \right) + K_D s$$

where K_P, K_I, and K_D are gains of proportional, integral, and derivative terms, respectively. It is used to improve stability and steady-state error.

PI controller: $G_c(s) = K_P + \left(K_I \middle/ s \right)$; it is used to improve the steady-state error.

PD controller: $G_c(s) = K_P + K_D s$; it is used to improve the stability of a system.

A PI and PD controller placed in a series is equivalent to a PID controller.

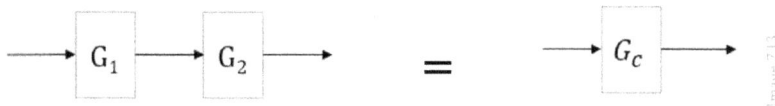

Let $G_1 = $ PI controller $= K_{P_1} + \left(K_I \middle/ s \right)$

Let $G_2 = $ PD controller $= K_{P_2} + K_D s$

$$G_c = G_1 G_2 = \left(K_{P_1} + K_I \middle/ s \right)(K_{P_2} + K_D s)$$

$$= (K_{P_1} K_{P_2} + K_I K_D) + \left(K_I K_{P_2} \middle/ s \right) + K_{P_1} K_D s$$

$$= (K_P)_e + \left[(K_I)_e \middle/ s \right] + (K_D)_e s$$

where the subscript, "e" represents the equivalent value.

Controller Gains K_P, K_I, and K_D

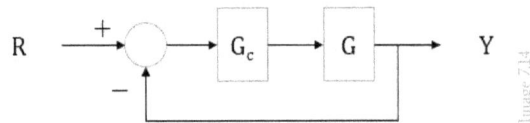

$$T = \frac{G_c G}{1 + G_c G}$$

$$G_c = K_P + \left(\frac{K_I}{s}\right) + K_D s$$

$$= \left(\frac{1}{s}\right)\left[K_P s + K_I + K_D s^2\right]$$

The controller gains can be calculated by the Zeigler-Nichols method:

$$K_P = 0.6 K_U, \; K_I = \frac{1.2 K_U}{T_U}, \; K_D = 0.075 K_U T_U$$

For a PI controller: $K_P = 0.45 K_U, \; K_I = 0.54 K_U / T_U$

For a PD controller: $K_P = 0.6 K_U, \; K_D = 0.075 K_U T_U$

For a proportional term controller: $K_P = 0.5 K_U$

where ultimate gain, $K_U = K$, for the system to be marginally stable, and it can be determined by the Routh-Hurwitz method. The ultimate period, $T_U = 2\pi/\omega$; $\omega =$ real root of characteristic equation, $q(s)$ with $s = j\omega$, evaluated at $K = K_U$.

Example

Given that $G(s) = \dfrac{1}{\left[s(s+1)(s+5)\right]}$

Find the PID controller's TF and the system's TF.

To find K_U and T_U, it is assumed that $G_c = K_P = K$.

The characteristic equation: $q(s) = 1 + G_c G = 0;\ 1 + KG = 0$

$$1 + \frac{K}{s(s+1)(s+5)} = 0$$

$$s(s+1)(s+5) + K = 0$$

$$s^3 + 6\,s^2 + 5\,s + K = 0$$

Routh Array

s^3	1	5	0
s^2	6	K	0
s^1	b_1	0	
s^0	K		

$$b_1 = (-K/6) + 5$$

For stability, b_1 and K should be positive.

For marginal stability, $b_1 = 0;\ (K/6) = 5.$

$$K = 30 = K_U$$

$$q(s) = s^3 + 6\,s^2 + 5\,s + K = 0$$

$$q(j\omega) = -\omega^2(j\omega) - 6\,\omega^2 + 5(j\omega) + 30 = 0$$

$$6(5 - \omega^2) + j\omega(5 - \omega^2) = 0; \ (5 - \omega^2)(6 + j\omega) = 0$$

$$\omega^2 = 5; \ \omega = \sqrt{5} \ \text{(Real value of } \omega)$$

$$T_U = 2\pi/\omega = 2\pi/\sqrt{5} = 2.81$$

$$K_P, \ K_I, \ K_D \ \text{for this example:}$$

$$K_P = 0.6K_U = 18$$

$$K_I = 1.2K_U/T_U = 12.81$$

$$K_D = 0.075K_U T_U = 6.32$$

$$G_c(s) = \left(\frac{1}{S}\right)\left[18S + 12.81 + 6.32 \ s^2\right]$$

$$= \left(\frac{6.32}{S}\right)\left[s^2 + 2.85 \ s + 2.03\right]$$

$$= \left(\frac{6.32}{S}\right)\left[(S + 1.425)^2\right]$$

ONE pole: $s = 0$

TWO zeros: $s = -1.425, \ s = -1.425$

Transfer Function

Controller TF: $G_c(s) = \left(\frac{6.32}{S}\right)\left(s^2 + 2.85 \ s + 2.03\right)$

$$G = \frac{1}{s(s + 1)(s + 5)}$$

$$G_c G = \frac{6.32(s^2 + 2.85 \ s + 2.03)}{s^2(s + 1)(s + 5)}$$

System TF:

$$T(S) = \frac{G_c G}{1 + G_c G} = \frac{6.32(s^2 + 2.85\ s + 2.03)}{s^2(s+1)(s+5) + 6.32(s^2 + 2.85\ s + 2.03)}$$

$$T(S) = \frac{6.3\ s^2 + 18\ s + 12.8}{s^4 + 6\ s^3 + 11.3\ s^2 + 18\ s + 12.8}$$

7.7. Summary

The poles of a closed-loop system are equal to the open-loop poles at $K = 0$ and the open-loop zeros at $K =$ infinity. So, the root locus diagram for a closed-loop system is based on the poles and zeros of the open-loop TF. When the controller gain (K) varies from zero to infinity, the loci travel from a pole to a zero. If a zero is not in the vicinity, a zero is assumed at infinity, and the loci will travel to infinity. Asymptotes guide their travel when it is going toward infinity. A locus exists in a region between two poles or between a pole and zero only if the controller gain is positive, and it does not exist if the controller gain is negative. The crossover points are the points where the locus or asymptote intersects the imaginary axis. The breakaway or break-in points are the points where the loci intersect the real axis. For complex poles and zeros, the departure angle from a complex pole and arrival angle at a complex zero must be determined.

7.8. Assessment

1. For $T = KP/(1 + KP)$, the poles of a closed-loop system at $K = 0$ are equal to:

   a. Open-loop poles

   b. Open-loop zeros

   c. Roots of the characteristic equation

   d. All of the above

2. The path or loci exist in a region where:

   a. $K > 0$

   b. $K = 0$

   c. $K < 0$

   d. None of the above

3. $K > 0$ in a region if the sum of poles and zeros to the right of an arbitrary point in this region is:

   a. Zero

   b. Odd

   c. Even

   d. None of the above

4. The value of K is zero at:

   a. A zero

   b. On the horizontal axis of the root locus

   c. On the vertical axis of the root locus

   d. A pole

5. At a given zero, the value of K is:

   a. Zero

   b. Infinity

   c. Positive

   d. Negative

6. If there is no path or loci in a region, then:

   a. $K = 0$

   b. $K = \infty$

   c. $K > 0$

   d. $K < 0$

7. If a locus travels to a zero at infinity, it needs:

    a. Crossover points

    b. A breakaway point

    c. Asymptotes

    d. A break-in point

8. Angle of departure is required for a locus to:

    a. Begin the travel from a real pole

    b. Begin the travel from a complex pole

    c. End the travel at a complex pole

    d. End the travel at a real pole

9. Crossover points for a locus are the points where:

    a. The system is marginally stable

    b. The locus intersects the vertical axis

    c. They represent the roots of the auxiliary polynomial

    d. All of the above

10. PID controllers are used to improve:

    a. Stability

    b. Steady-state error

    c. Control of industrial processes

    d. All of the above

7.9. Practice Problems

1. Answer all the assessment questions in section 7.8

2. A typical control system with controller gain K, and Plant transfer function, $P(s)$ is given below along with the system transfer function, $T(s)$.

$$T(s) = \frac{K(s+5)}{(s^2 + 4s + 3) + K(s+5)}$$

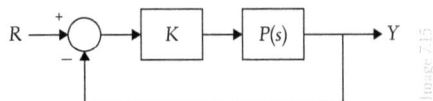

a. Determine the Plant transfer function, $P(s)$ and draw the root locus diagram.

b. Determine the break-away and break-in points and respective values of gain, K.

3. A typical control system with controller gain K, and Plant transfer function, $P(s)$ is given below.

$$P(s) = \frac{(s+1)}{(s^2 + 4s + 5)}$$

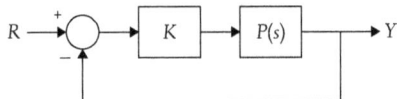

a. Determine the break-in point for the loci and the departure angle for complex pole

b. Draw the root locus diagram and show all the values on the diagram.

4. A typical control system with controller gain K, and Plant transfer function, $P(s)$ is given below.

$$P(s) = \frac{1}{s(s^2 + 2s + 5)}$$

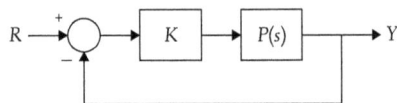

a. Determine asymptotes, angle of departure, loci cross-over points and asymptote cross-over points

b. Draw the root locus diagram

5. A typical control system with controller gain K, and Plant transfer function, $P(s)$ is given below.

$$P(s) = \frac{1}{s(s^2 + 7s + 10)}$$

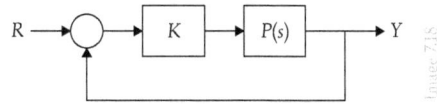

R → ◯ → | K | → | P(s) | → Y

a. Determine the break-away point and the value of K at this point.

b. Draw the root locus diagram and show the cross-over points.

6. For given ultimate gain, $K_U = 25$ and ultimate period, $T_U = 2.5$ sec, determine the controller transfer function for a (a) PI controller and (b) PD controller.

7. For given ultimate gain, $K_U = 468$ and ultimate period, $T_U = 1.05$ sec, determine the transfer function for a PID controller.

8. For given ultimate gain, $K_U = 45$ and ultimate period, $T_U = 5$ sec, determine the equivalent controller transfer function when a PI controller and a PD controller are connected in series as shown below.

$$G_1(s) = K_{P1} + \frac{K_1}{(s)}; \ G_2(s) = K_{P2} + K_D(s)$$

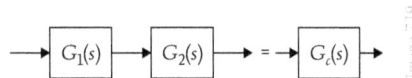

→ | G₁(s) | → | G₂(s) | → = → | G_c(s) | →

9. The equivalent controller transfer function for a system with PI and PD controllers connected in series is given below.

$$G_c(s) = 124.5 + \frac{60}{(s)} + 22.5(s)$$

Determine the values of K_{P1}, K_{P2}, K_1 and K_D assuming $K_{P1} = (3/4) K_{P2}$

10. Determine the values of ultimate gain, *Ku* and ultimate period *Tu* for the feedback system given below.

$$P(s) = \frac{1}{s(s^2 + 10s + 16)}$$

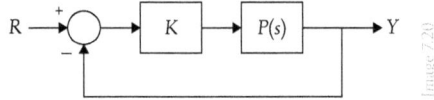

R → ⊕ → [K] → [P(s)] → Y

Image 7.20

MATLAB problems:

11. Draw the root locus diagram using MATLAB for the plant transfer function given below:

a. $P(s) = \dfrac{25}{(s^3 + 10s^2 + 40s + 25)}$

b. $P(s) = \dfrac{(s + 10)}{(s^2 + 2s + 10)}$

c. $P(s) = \dfrac{(s^2 + 2s + 4)}{s(s^2 + 5s + 10)}$

d. $P(s) = \dfrac{(s^5 + 6s^4 + 6s^3 + 12s^2 + 6s + 4)}{(s^6 + 4s^5 + 5s^4 + s^3 + s^2 + 12s + 1)}$

12. The control system of a typical mechanical system is given below. Draw the root locus diagram using MATLAB.

a. $P(s) = \dfrac{10(s + 1)}{s(s^2 + 4.5s + 9)}$

b. $P(s) = \dfrac{(s^2 + 4s + 8)}{s(s + 2)}$

c. $P(s) = \dfrac{(s^2 + 4s + 8)}{(s^2 + 2s + 2)}$

SIMULINK problems:

13. For the control system given below, determine the response using SIMULINK for step input when,

    a. $K = 0.435$

    b. $K = 1$

    c. $K = 1.51$

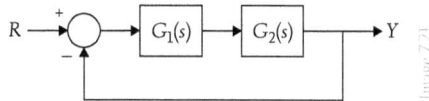

$$G_1(s) = \frac{K(s+1)}{s}; \quad G_2(s) = \frac{10}{s^2 + 4.5s + 9}$$

14. A negative feedback control system has a PID controller as shown below with $K_P = 54$, $K_1 = 51.4$, and $K_D = 14.2$

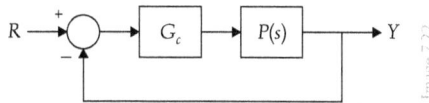

The plant transfer function, $P(s) = P(s) = \dfrac{1}{s(s+1)(s+9)}$

Determine the response using SIMULINK for the following input,

    a. Step input

    b. Ramp input

    c. Impulse input

    d. Sinusoidal input

BODE AND NYQUIST PLOTS

8.1. Introduction

The transfer function (TF) can be expressed in a frequency domain. Then, its magnitude in decibel and phase angle in degrees can be obtained as a function of frequency. They can be represented in two separate plots with a semilog scale as a Bode plot or together on polar graph paper as a Nyquist plot. This chapter also discusses the system bandwidth frequency, phase margin, and gain margin. It also demonstrates how to determine the damping ratio from the phase margin.

8.2. Learning Objectives

1. Express the magnitude of a system's TF in a frequency domain in terms of decibels (dB).

2. Draw the Bode plots of magnitude (dB) and phase angle in separate plots as a function of frequency.

3. Draw the Nyquist plot of magnitude and phase angle in a same plot (polar plot).

4. Determine gain margin, phase margin, and bandwidth frequency.

8.3. Frequency Diagrams

Hendrik Wade Bode (1905–1982) invented Bode plots to represent the magnitude and phase angle of a system's TF in two separate logarithmic graphs. He also invented gain margin, phase margin, and bandwidth in the stability analysis of linear systems. Harry Theodor Nyquist (1889–1976) represented the magnitude and phase angle together in one plot using a polar graph. These graphs are called *Nyquist plots* and provide an alternate method in stability analysis.

8.3.1. Transfer Function as a Function of Frequency

Let the system $TF = T(s)$. It can be expressed as a function of frequency by substituting $s = j\omega$. Then, the TF $T(j\omega)$ would have a ratio of complex values. The magnitude of TF is expressed in terms of dBs, and the phase angle is expressed in terms of degrees.

DECIBEL

A *bel* is defined as the logarithmic ratio of two power levels. A dB is equal to one-tenth of a bel, and it is commonly used in the measurement of noise in acoustic analysis.

Sound level: $L = \log_{10}\left(\dfrac{P_1}{P_2}\right)$ Bel; 1Bel = 10dB; $L = 10\left[\log_{10}\left(\dfrac{P_1}{P_2}\right)\right]$ dB

where P_1 and P_2 are power levels.

Power: $P = \omega T$; torque: $T = I\alpha$ where 'I' is the mass moment of inertia and $\alpha = \left(\dfrac{\omega}{t}\right)$.

So, $P = \omega\left(\dfrac{I\omega}{t}\right) = \dfrac{I}{t}\omega^2$:

$$\frac{P_1}{P_2} = \frac{\omega_1^2}{\omega_2^2}$$

$$L = 10\log_{10}\left(\frac{\omega_1}{\omega_2}\right)^2 = 20\log_{10}\left|\frac{\omega_1}{\omega_2}\right| \, dB$$

For an octave: $\omega_2 = 2\omega_1$; $L = \left|20\log(1/2)\right| = 6dB$

For a decade: $\omega_2 = 10\omega_1$; $L = \left|20\log(1/10)\right| = 20dB$

For comparison of dB level, the auditory threshold is 0 dB, a whisper is 30 dB, loud thunder is 110 dB, a jet engine's noise during the takeoff of an airplane is 140 dB, and a rocket engine's noise during takeoff is 180 dB.

Similarly for a given TF, $T(s)$, substitute $s = j\omega$ to find $T(j\omega)$. Then, the magnitude in dB = 20 log $|T(j\omega)|$; phase angle, $\phi == \angle T(j\omega)$.

8.3.2. Magnitude and Phase Angle

CASE (I)

If $T(j\omega) = (A + jB)^N$; $\left|T\right| = (A^2 + B^2)^{N/2}$

Magnitude: $dB = 20\log\left|T\right| = 20(N/2)\log(A^2 + B^2)$

$$= (10N)\log(A^2 + B^2)$$

Phase angle: $\phi = (N) \tan^{-1}(B/A)$

$$A = 0; \, dB = 20N \log(B), \, \phi = (N)(\pi/2)$$

$$B = 0; \, dB = 20N \log(A), \, \phi = 0$$

Log algebra:

$$\log(ab) = \log(a) + \log(b)$$

$$\log(a/b) = \log(a) - \log(b)$$

CASE (II)

$$T(j\omega) = \frac{K(z + j\omega)}{(j\omega)^2(a + j\omega)(b + j\omega)} = N/D$$

Magnitude: $dB = 20 \log|T| = 20 \log|N| - 20 \log|D|$

$$= 20 \log K + 10 \log(z^2 + \omega^2) - 20(2) \log \omega - 10 \log(a^2 + \omega^2) - 10 \log(b^2 + \omega^2)$$

Phase angle:

$$\phi = \angle T = \angle N - \angle D = 0 + \tan^{-1}(\omega/z) - 2(\pi/2) - \tan^{-1}(\omega/a) - \tan^{-1}(\omega/b)$$

CASE (III)
A typical TF, $T(s)$, can be expressed as:

$$T(s) = \frac{K(1 + \tau_1 s)}{s^N(1 + \tau_2 s)(s^2 + 2\zeta\omega_n s + \omega_n^2)}$$

$$T(j\omega) = \frac{K(1 + j\omega\tau_1)}{(j\omega)^N(1 + j\omega\tau_2)(-\omega^2 + j(2\zeta\omega_n)\omega + \omega_n^2)}$$

$$= \frac{K_1(1 + j\omega\tau_1)}{(j\omega)^N(1 + j\omega\tau_2)\{(1 - u^2) + j(2\zeta u)\}}$$

where $K_1 = (K/\omega_n^2)$ and $u = (\omega/\omega_n)$:

$$dB = 20\log\left|T(j\omega)\right|$$

$$= \left[20\log K_1 + 10\log(1 + \omega^2\tau_1^2)\right] - \left[20N\log\omega + 10\log(1 + \omega^2\tau_2^2) + 10\log\left\{(1 - u^2)^2 + (2\zeta u)^2\right\}\right]$$

$$\phi = \tan^{-1}(\omega\tau_1) - N(\pi/2) - \tan^{-1}(\omega\tau_2) - \tan^{-1}\left[2\zeta u/(1 - u^2)\right]$$

EXAMPLE

$$T(s) = 1000 / \left[(s + 10)(s + 100)\right]$$

Find the magnitude in dB and phase angle in degrees for $\omega = 0, 10, 200$ rad/s.

$$T(j\omega) = 1000/\left[(10 + j\omega)(100 + j\omega)\right]$$

Magnitude: $dB = 20\log\left|T(j\omega)\right|$

$$= 20\log(1000) - \left[10\log(100 + \omega^2) + 10\log(10000 + \omega^2)\right]$$

$$= 60 - \left[10\log(100 + \omega^2) + 10\log(10000 + \omega^2)\right]$$

Phase angle: $\phi = = \angle T(j\omega)$

$$\phi = 0 - \tan^{-1}(\omega/10) - \tan^{-1}(\omega/100)$$

ω	dB	ϕ(deg)
0	0	0
10	-3.05	-50.71
200	-33.02	-150.57

8.3.3. Peak Value of Magnitude and Frequency Ratio

$$\text{For } G(j\omega) = \left(1 \Big/ \left[(1-u^2) + j(2\zeta u)\right]\right); \; u = \left(\omega \Big/ \omega_n\right)$$

$$\left|G(j\omega)\right| = \left(1 \Big/ \sqrt{(1-u^2)^2 + (2\zeta u)^2}\right) = \left(1 \Big/ \sqrt{\left\{u^4 + 2u^2(2\zeta^2 - 1) + 1\right\}}\right) = \left(1 \Big/ \sqrt{\beta}\right)$$

$$\beta = u^4 + 2u^2(2\zeta^2 - 1) + 1$$

Peak value of magnitude: $M_{p\omega} = \left|G\right|_{\max}$; $(dB)_{\max} = 20\log\left|G\right|_{\max} = 20\log M_{p\omega}$

For $\left|G\right| = \left|G\right|_{\max}, \; \dfrac{d\left|G\right|}{du} = 0$

$$\frac{d\left|G\right|}{du} = \frac{-\left\{4u^3 + 4u(2\zeta^2 - 1)\right\}}{\left\{u^4 + 2u^2(2\zeta^2 - 1) + 1\right\}} = 0$$

$$4u^3 + 4u(2\zeta^2 - 1) = 0$$

$$u^2 = 1 - 2\zeta^2$$

$$u = \sqrt{1 - 2\zeta^2} = u_r = \frac{\omega_r}{\omega_n}; \; \omega = \omega_r \; @ \left|G\right| = \left|G\right|_{\max}$$

Substitute this value of u in $\left|G\right|$:

$$\left|G\right|_{\max} = \left(1 \Big/ \sqrt{\beta_p}\right) = M_{p\omega}$$

$$\beta_p = (1 - 2\zeta^2)^2 - 2(1 - 2\zeta^2)^2 + 1 = 1 - (1 - 2\zeta^2)^2 = 4\zeta^2(1 - \zeta^2)$$

PEAK VALUE

$$\text{Magnitude: } M_{p\omega} = \left(\frac{1}{\sqrt{\beta_p}}\right) = \left[\frac{1}{\left(2\zeta\sqrt{1-\zeta^2}\right)}\right]$$

$$u_r = \frac{\omega_r}{\omega_n} = \sqrt{1-2\zeta^2}; \text{ Frequency, } \omega_r = \omega_n\sqrt{1-2\zeta^2}$$

For $\zeta < 1$

ζ	$M_{p\omega}$	ω_r/ω_n
0	∞	1
0.1	5.025	0.99
0.2	2.55	0.96
0.707	1.0	0

EXAMPLE

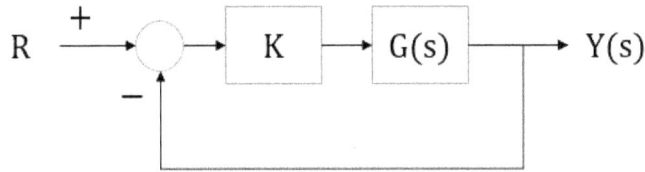

For $G = \dfrac{1}{s(s+1)(s+2)}$ and $K = 2$, draw the Bode plot.

$$T(s) = {KG}\Big/{(1 + KG)}$$

$$KG = \frac{2}{s(s+1)(s+2)}; T(s) = \frac{2}{s(s+1)(s+2)+2} = \frac{2}{(s^3 + 3s^2 + 2s + 2)}$$

$$T(j\omega) = \frac{2}{\left[(-\omega^2) = j\omega - 3\omega^2 + 2(j\omega) + 2\right]} = \frac{2}{\left[(2 - 3\omega^2) + j\omega(2 - \omega^2)\right]}$$

Magnitude: $dB = 20\log 2 - 10\log\left\{(2 - 3\omega^2)^2 + \omega^2(2 - \omega^2)^2\right\}$

Phase angle: $\phi = 0 - \tan^{-1}\dfrac{\omega(2 - \omega^2)}{(2 - 3\omega^2)}$

ω	dB	ϕ(deg)
0.1	0.09	-6
0.8	5	-86
1.0	3	-135
2.0	-14.6	-202

EXAMPLE

For the same example above, find $M_{p\omega}, \zeta,$ percent overshoot (PO), T_s.

$$(dB)_{max} = 5;\ \omega_r = 0.8\ \left(frequency\ at\ dB = (dB)_{max}\right)$$

$$(dB)_{max} = 20\log\left|T\right|_{max} = 20\log M_{p\omega} = 5\ dB;\ \log M_{p\omega} = 5/20 = 0.25$$

$$M_{p\omega} = (10)^{0.25} = 1.8$$

$$M_{p\omega} = \frac{1}{2\zeta\sqrt{1 - \zeta^2}} = 1.8$$

$$\zeta^4 - \zeta^2 + 0.077 = 0$$

$$\zeta = 0.29,\ 0.96$$

Choose $\zeta = 0.29$ because $\zeta < 1$; the other value is close to one.

The maximum value of 5 dB occurs at $\omega = 0.8$; $\omega_r = 0.8$.

$$\omega_n = \frac{\omega_r}{\sqrt{1-2\zeta^2}} = \frac{0.8}{\sqrt{1-2(0.29)^2}} = 0.88$$

$$PO = 100e^{\left(-\zeta\pi/\sqrt{1-\zeta^2}\right)} = 38.6\%$$

$$T_s = \frac{4}{\zeta\omega_n} = 15.67\,\text{sec}$$

8.3.4. Bandwidth Frequency (ω_B)

Bandwidth frequency is defined as the frequency at which the magnitude is equal to -3dB. Because one octave is 6 dB, the bandwidth frequency is also defined as the frequency at which the frequency response has declined one-half octave from 0 dB.

EXAMPLE

$$T(s) = \frac{120}{s(s+1)(s+2)}$$

Find the bandwidth frequency, ω_B, and the respective phase angle, ϕ_B.

$$T(j\omega) = \frac{120}{(j\omega)(j\omega+1)(j\omega+2)}$$

$$dB = 20\log|T| = 20\log(120) - 20\log(\omega) - 10\log(\omega^2+1) - 10\log(\omega^2+4)$$

$$\phi = 0 - \left(\pi/2\right) - \tan^{-1}(\omega) - \tan^{-1}\left(\omega/2\right)$$

ω	dB	ϕ(deg)
3	10.9	-217.9
4	4.25	-229.4
5	-1.15	-236.9
6	-5.66	-242
7	-9.53	-246

dB	ω
−1.15	5
−3	?
−5.66	6

LINEAR INTERPOLATION

To find the frequency (ω) at given magnitude (dB), linear interpolation is used.

$$\frac{\omega - \omega_1}{dB - dB_1} = \frac{\omega_2 - \omega_1}{dB_2 - dB_1}$$

where dB_1 and dB_2 are known values of magnitude at respective frequencies of ω_1 and ω_2. For given value of $dB = -3$ such that $\left|dB_1\right| < \left|dB\right| < \left|dB_2\right|$, the bandwidth frequency ω_B can be found by interpolating for ω as shown below.

$$\omega = \omega_1 + \left[(dB - dB_1)\frac{(\omega_2 - \omega_1)}{(dB_2 - dB_1)}\right]; \qquad \omega = \omega_B \text{ at } dB = -3$$

$$\omega_B = 5 + \left[(-1.85)(1)/(-4.51)\right] = 5.4$$

Bandwidth frequency: $\omega_B = 5.4$ rad/sec

Phase angle: $\phi_B = \phi$ at $\omega = \omega_B$

$$= -90 - \tan^{-1}(5.4) - \tan^{-1}(2.7) = -90 - 79.5 - 69.7$$

$$\phi_B = -239°$$

SECOND-ORDER SYSTEM: BANDWIDTH FREQUENCY AND DAMPING RATIO

For a second-order system:

$$T(s) = \frac{\omega_n^2}{s^2 + (2\zeta\omega_n)s + \omega_n^2}; \zeta < 1; T(j\omega_B) = \frac{\omega_n^2}{\left(\omega_n^2 - \omega_B^2\right) + j2\xi\omega_n\omega_B}$$

$$= \frac{1}{(1 - u^2) + i(2\zeta u)}; \text{ where } u = \frac{\omega_B}{\omega_n}; \quad |T| = \frac{1}{\sqrt{(1 - u^2)^2 + (2\zeta u)^2}}$$

At $20\log|T| = -3$ dB, $\log|T| = -\left(\frac{3}{20}\right); |T| = 0.7079 = \frac{1}{\sqrt{2}}$

$$(1 - u^2)^2 + 4\zeta^2 u^2 = 2; \quad u^4 + (4\zeta^2 - 2)u^2 - 1 = 0$$

$$u^2 = (1 - 2\zeta^2) + \sqrt{(2\zeta^2 - 1)^2 + 1}$$

$$u = \left[(1 - 2\zeta^2) + \sqrt{4\zeta^4 - 4\zeta^2 + 2}\right]^{(1/2)} = \frac{\omega_B}{\omega_n}$$

$$\omega_B = \omega_n \left[(1 - 2\zeta^2) + \sqrt{4\zeta^4 - 4\zeta^2 + 2}\right]^{(1/2)} = f(\omega_n, \zeta)$$

For a given PO and settling time, the value of ζ can be determined from the PO, while the value of ω_n can be determined from the settling time. Then, the ω_B can be easily determined from the above equation.

EXAMPLE

Given $T(s) = \dfrac{81}{s^2 + 9s + 81}$:

Find $\omega_B, M_{p\omega}$, and ω_r.

Characteristic equation: $s^2 + 9s + 81 = 0$

Compare with $s^2 + (2\zeta\omega_n)s + \omega_n^2 = 0$:

$$\omega_n^2 = 81; \omega_n = 9$$

$$2\zeta\omega_n = 9; \zeta = 0.5$$

Bandwidth frequency: $\omega_B = \omega_n \left[(1 - 2\zeta^2) + \sqrt{4\zeta^4 - 4\zeta^2 + 2}\right]^{(1/2)} = 11.45$

Peak value of magnitude: $M_{p\omega} = \dfrac{1}{2\zeta\sqrt{1-\zeta^2}} = 1.15$

Peak value frequency: $\omega_r = \omega_n\sqrt{1-2\zeta^2} = 6.36$

8.4. Nyquist Stability Criterion

It is based on mapping contours from an S-plane to an F(s)-plane.

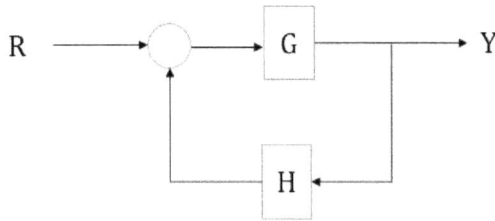

$$T(s) = \frac{G}{1+GH} = \frac{G(s)}{F(s)}$$

Characteristic equation: $F(s) = 1 + GH = 1 + L(s)$; $L(s) = G(s)H(s) = \dfrac{N(s)}{D(s)}$

$$F(s) = \frac{D+N}{D}$$

The poles of F(s) are same as the poles of L(s). A Nyquist plot is a polar plot of magnitude and angle of $L(j\omega)$. $L(s) = F(s) - 1$; $L(s) = -1$ for $F(s) = 0$. Then, T(s) becomes infinity, leading the system to be unstable. So for stability, $L(s) < (-1)$.

The Nyquist stability criterion states that a feedback system is stable if and only if the contour of the plot does not encircle the $(-1,0)$ point.

EXAMPLE

Sketch a Nyquist plot for a given $L(s) = \dfrac{1000}{(s+1)(s+10)}$.

$$L(s) = \frac{1000}{s^2 + 11s + 10}$$

$$L(j\omega) = \frac{1000}{\left[(10 - \omega^2) + j(11\omega)\right]}$$

$$\left|L(j\omega)\right| = \frac{1000}{\sqrt{(10 - \omega^2)^2 + 121\omega^2}}$$

$$\angle L(j\omega) = 0 - \tan^{-1}\left(\frac{11\omega}{10 - \omega^2}\right)$$

POLAR PLOT OF $|L|$ AND $\angle L$

$$L(j\omega) = f(r,\phi) \text{ where } r = |L| \text{ and } \phi = \angle L$$

ω	r	$\phi(\text{deg})$
0	100	0
1	70.35	−50.71
10	6.8	−129.3
100	0.1	−173.7
∞	0	−180

The negative sign represents that the angles are measured in clockwise direction. The plot is symmetrical. MATLAB can be used to generate the plot. The procedure for a manual plot is given below.

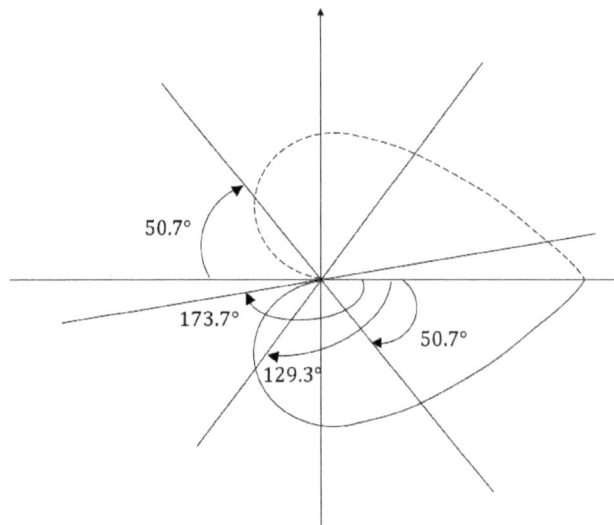

The contour does not encircle the point $(-1,0)$, and hence the system is stable.

EXAMPLE

$$L(s) = \frac{K}{s(s+1)^2} = \frac{K}{s^3 + 2s^2 + s}$$

$$L(j\omega) = \frac{K}{(-2\omega^2) + j(\omega - \omega^3)}$$

Magnitude: $\left|L(j\omega)\right| = \dfrac{K}{\sqrt{4\omega^4 + (\omega - \omega^3)^2}}$

Phase angle: $\phi = \angle L(j\omega) = 0 - \tan^{-1}\left\{\dfrac{\omega\left(1 - \omega^2\right)}{-2\omega^2}\right\} = \tan^{-1}\left(\dfrac{1 - \omega^2}{2\omega}\right)$

The Nyquist plots for different values of K are provided below.

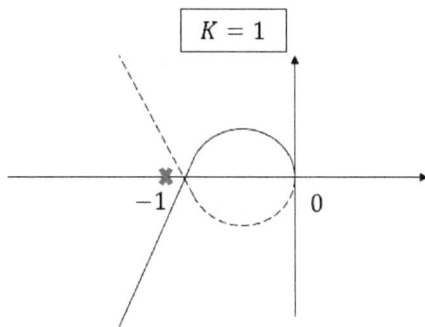

K = 1

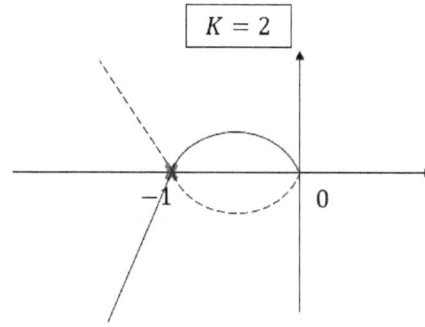

K = 2

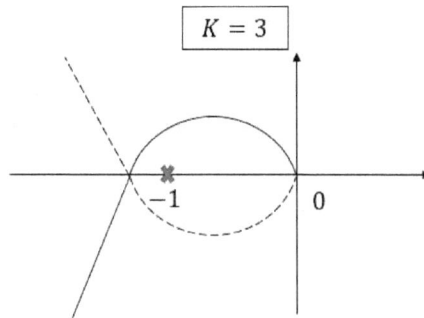

K = 3

The magnitude of $L(j\omega)$ at $\angle L = -180°$ can be determined by equating:

$$\tan^{-1}\left(\frac{1-\omega^2}{2\omega}\right) = -180°$$

$$\frac{(1-\omega^2)}{2\omega} = \tan(-180) = 0; \ \omega = 1$$

Then, $L(j\omega) = K/(-2)$.

At $K = 1$, $L = -(1/2)$; the contour does not encircle the point $(-1,0)$, and the system is stable.

$K = 2$, $L = -1$; the contour cuts through the point $(-1, 0)$, and the system is marginally stable.

$K = 3$, $L = -1.5$; the contour encircles the point $(-1,0)$, and the system is unstable.

8.5. Relative Stability

In the time domain, a system with a lower value of settling time (T_s) is considered relatively more stable than a system with a higher value of settling time. In the frequency domain, the parameters used to measure relative stability are gain margin (GM) and phase margin (PM).

The GM is defined as the increase in system gain at the phase crossover frequency, ω_0 (when the phase angle, $\phi = -180°$), that will result in a marginally stable system. From Bode plot, the GM can be obtained as THE magnitude in dB at the phase crossover frequency, ω_0. It is equivalent to the intersection of the $(-1,0)$ point on the Nyquist diagram.

$$\omega = \omega_0 \text{ at } \phi = -180°.$$

$GM = 20\log|L|$, evaluated at the phase cross-over frequency, ω_0.

The PM is defined as the amount of phase shift at the gain crossover frequency, ω_c (when the magnitude is zero dB or $|L| = 1$), that will result in a marginally stable system. From a Bode plot, the PM can be obtained as the phase angle in degrees at the gain crossover frequency, ω_c. It is equivalent to the intersection of the $(-1,0)$ point on the Nyquist diagram.

$\omega = \omega_c$ at $|L| = 1$; $PM = \angle L$, evaluated at the gain crossover frequency, ω_c.

A typical Bode plot with the PM and GM is provided below.

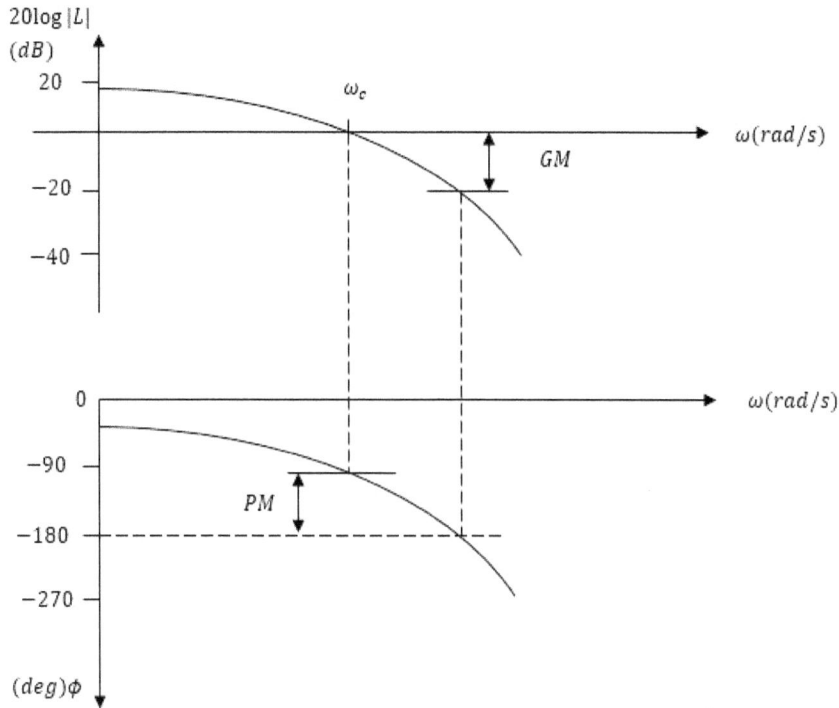

8.5.1. Computation of GM and PM

Given the TF, $L(s)$, determine $L(j\omega)$ and then find the magnitude in dB $\left(20\log|L(j\omega)|\right)$ and the angle, $\angle L$.

<u>GM</u> (dB): Set $\angle L = -180$ and find $\omega = \omega_0$, the phase crossover frequency.

Then GM $= 20\log|L(j\omega)|$ evaluated at ω_0.

<u>PM</u> (degrees): Set $|L(j\omega)| = 1$ and find $\omega = \omega_c$, gain crossover frequency.

Then, PM $= \angle L$ evaluated at ω_c.

EXAMPLE

$$L(s) = \frac{K}{(s+2)(s+3)}$$

a. Find the GM for $K = 1$.

b. Find K for given GM of -3 dB.

c. Find PM for $K = 10$.

$$L(s) = \frac{K}{s^2 + 5\,s + 6}$$

$$L(j\omega) = \frac{K}{\left[(6 - \omega^2) + j(5\,\omega)\right]}$$

$$|L| = \frac{K}{\sqrt{(6 - \omega^2)^2 + 25\,\omega^2}}; \quad \angle L = -\tan^{-1}\frac{5\,\omega}{(6 - \omega^2)}$$

$$(6 - \omega^2)^2 + 25\,\omega^2 = \omega^4 + 13\,\omega^2 + 36 = (\omega^2 + 4)(\omega^2 + 9)$$

$$|L| = \frac{K}{\left[(\omega^2 + 4)(\omega^2 + 9)\right]^{(1/2)}}$$

$$dB = 20\log|L| = 20\log K - 10\log(\omega^2 + 4) - 10\log(\omega^2 + 9)$$

a. $K = 1$

$$\angle L = -180; \, -\tan^{-1}\frac{5\,\omega}{(6 - \omega^2)} = -180; \, \frac{5\,\omega}{(6 - \omega^2)} = \tan(180) = 0; \, \omega_0 = 0$$

$$dB = 0 - 10\log 4 - 10\log 9 = -15.56$$

$$GM = -15.56 \text{ dB}$$

b. $GM = -3$ dB

$$GM = 20\log K - 10\log 4 - 10\log 9$$

$$-3 = 20\log K - 15.56$$

$$20\log K = 12.56; \, K = 4.246$$

c. $K = 10$

$$|L| = \frac{K}{[\omega^4 + 13\omega^2 + 36]^{(1/2)}} = 1$$

$$K^2 = \omega^4 + 13\omega^2 + 36 = 100$$

$$\omega^4 + 13\omega^2 - 64 = 0$$

$$\omega^2 = \left(\frac{1}{2}\right)\left[-13 \pm \sqrt{425}\right] = 3.8, -16.8$$

Because ω is positive, $\omega_c = \sqrt{3.8} = 1.95$:

$$\angle L = -\tan^{-1}\frac{5\omega_c}{(6 - \omega_c^2)} = -77°$$

$$PM = -77°$$

8.5.2. Phase Margin and Damping

For a second-order system, $L(s) = \dfrac{\omega_n^2}{s(s + 2\zeta\omega_n)}$:

$$L(j\omega) = \frac{\omega_n^2}{(-\omega^2) + j(2\zeta\omega_n\omega)}$$

$$|L| = \frac{\omega_n^2}{\left[\omega^4 + \left(4\zeta^2\omega_n^2\right)\omega^2\right]^{(1/2)}}$$

$$\angle L = -\tan^{-1}\left(\frac{2\zeta\omega_n\omega}{-\omega^2}\right) = \tan^{-1}\left(\frac{2\zeta\omega_n}{\omega}\right)$$

$$\omega = \omega_c \text{ at } |L| = 1$$

$$\omega^4 + (4\zeta^2\omega_n^2)\omega^2 - \omega_n^4 = 0$$

Solving:

$$\omega^2 = \omega_n^2 \left[-2\zeta^2 \pm \sqrt{4\zeta^4 + 1} \right]$$

$$\omega_c = \omega_n \left[-2\zeta^2 + \sqrt{4\zeta^4 + 1} \right]^{(1/2)}$$

$$\text{PM} = \tan^{-1}\left(\frac{2\zeta\omega_n}{\omega_c} \right) = \tan^{-1}\left\{ \frac{2\zeta}{\left[-2\zeta^2 + \sqrt{4\zeta^4 + 1} \right]^{(1/2)}} \right\}$$

Linear approximation: $\text{PM} = 100\zeta$

$$\zeta = 0.01(\text{PM}) = 0.01\phi$$

This approximation is good for $10° \le \phi \le 70°$.

It results in a damping range of $0.1 \le \zeta \le 0.707$.

8.6. Summary

The Bode plot represents the magnitude of the TF in dBs and the phase angle in degrees on a semilog graph sheet. To obtain this, the TF is first expressed in terms of frequency by replacing the Laplace parameter, S, by $j\omega$. Then, it is shown that the magnitude, $\text{dB} = 20 \log|T|$, and the phase angle, $\phi = \angle T(j\omega)$. A method is presented to determine the bandwidth frequency when the magnitude is equal to -3 dB. In a Nyquist plot, the magnitude and phase angle are represented in a single polar plot. The phase margin and gain margin based on a Bode plot are useful to study the relative stability of a control system. For a second-order system, the damping ratio is approximately 1% of its phase margin.

8.7. Assessment

1. A Bode plot is:

    a. A logarithmic plot

    b. A polar plot

    c. A linear plot

    d. None of the above

2. A Nyquist plot is:

    a. A logarithmic plot

    b. A polar plot

    c. A linear plot

    d. None of the above

3. If $T = \dfrac{1}{3 + j4}$, the magnitude $|T|$ is:

    a. 1/5

    b. 1/7

    c. 1/25

    d. 25

4. The angle, $\angle T$ is:

    a. $37°$

    b. $-37°$

    c. $-53°$

    d. $53°$

5. If $T = 4, \angle T$ is:

    a. $90°$

    b. $0°$

    c. $76°$

    d. $0.07°$

6. If $T = j4$, $\angle T$ is:

   a. $90°$

   b. $0°$

   c. $76°$

   d. $0.07°$

7. Bandwidth frequency is the frequency at:

   a. Magnitude $= -3$ dB

   b. Magnitude $= -\left(\dfrac{1}{2}\right)$(octave)

   c. $|T| = 0.71$

   d. All of the above

8. If $|T| = -1$ dB @ $w = 5$ and -5 dB @ $w = 6$, the bandwidth frequency is:

   a. 11

   b. 5/6

   c. 5.5

   d. 6/5

9. The phase margin is the phase angle evaluated at:

   a. The gain crossover frequency

   b. Magnitude is 0 dB

   c. Magnitude $= 1$

   d. All of the above

10. The gain margin is the magnitude in dB evaluated at:

   a. Frequency at $\phi = -180°$

   b. Frequency at $\phi = 90°$

   c. Frequency at $\phi = -90°$

   d. Frequency at $\phi = 180°$

8.8. Practice Problems

1. Answer all the assessment questions in Section 8.7

2. Determine the magnitude in dB and phase angle in degrees at $\omega = 10, 20, 30$ for the transfer function given below.

$$T(s) = \frac{100(s+5)^8}{s^2(s+3)(s+2)}$$

3. Determine the magnitude in dB and phase angle in degrees at $\omega = 5$ for the transfer function given below.

$$T(s) = \frac{50(1+4s)}{s^{10}(1+6s)(s^2+6s+25)}$$

4. Determine the magnitude in dB and phase angle in degrees at $\omega = 0, 0.5, 1, 2, 4, 10, 100$ for the transfer function given below.

$$L(s) = \frac{4}{(s+2)^2}$$

   a. Draw the Bode plots on a semi-Log sheet with frequency on log scale in horizontal axis while the vertical axis is on regular scale.

   b. Draw the Bode plots using MATLAB

5. Determine the bandwidth frequency for the transfer function given below.

$$T(s) = \frac{25}{(s^2+3s+25)}$$

6. Determine the bandwidth frequency and the respective phase angle for the transfer function given below.

$$T(s) = \frac{160}{s(s^2+9s+20)}$$

7. Determine the magnitude and phase angle for the Nyquist plot at $\omega = 0, 1, 10, 100, \infty$ when the transfer function is given as,

$$L(s) = \frac{1500}{(s^2 + 14s + 24)}$$

Draw the Nyquist plot using MATLAB

8. Determine the values of "K" for the transfer function given below such that the system is (i) stable, (ii) marginally stable, and (iii) unstable, based on Nyquist theory.

$$L(s) = \frac{K}{s(s^2 + 3s + 9)}$$

9. Determine the Phase Margin and Gain Margin for the transfer function given below.

$$L(s) = \frac{15}{(s + 3)(s + 4)}$$

10. Determine the natural frequency and damping ratio for the transfer function given below. Based on these values, determine the gain cross-over frequency and Phase Margin.

$$L(s) = \frac{16}{(s^2 + 4s + 16)}$$

CONTROL SYSTEM DESIGN

9.1. Introduction

Control system design involves minimizing performance deficiency. The component used for the purpose of compensating for this deficiency is called a *compensator*. It may be used for amplifying with a phase lead or attenuating with a phase lag. This chapter discusses the design of a phase lead compensator and a phase lag compensator. However, the zero in the compensator's transfer function (TF) affects the total system's TF. To eliminate this unwanted effect, a prefilter is required, and its design is also included in this chapter. In general, the compensators are the P-I type of compensators. This chapter demonstrates how to design a prefilter with a P-I compensator.

9.2. Learning Objectives

1. Design a lead compensator.

2. Design a prefilter.

3. Design a lag compensator.

9.3. Compensator Design

A compensator (G_c) is an additional component added to a control system to compensate for performance deficiency. There are four types of compensators based on where the compensator is added in the loop of control system.

i.   <u>Cascade type</u>: The compensator is added to the feed-forward loop, as shown below.

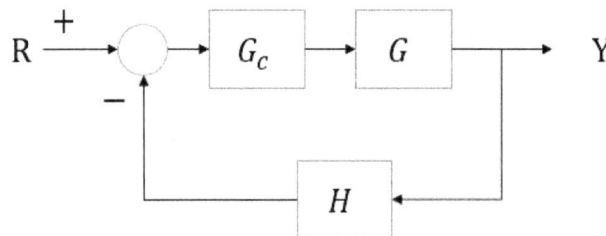

$G_c \rightarrow$ Compensator TF

$G \rightarrow$ Plant TF

ii. __Feedback type:__ The compensator is added to the feedback loop, as shown below.

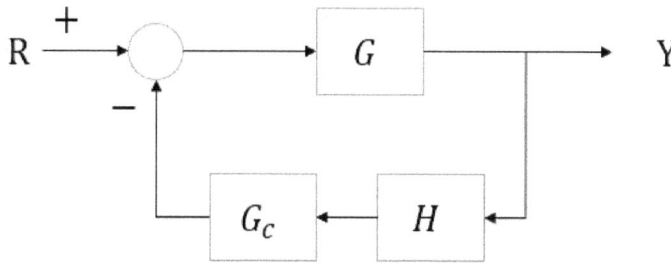

iii. __Output type:__ The compensator is added to the output part of the loop after the pick-off point, as shown below.

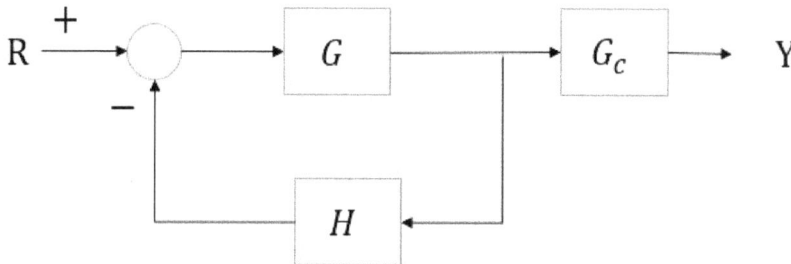

iv. __Input type:__ The compensator is added to the input part of the loop before the summing point, as shown below.

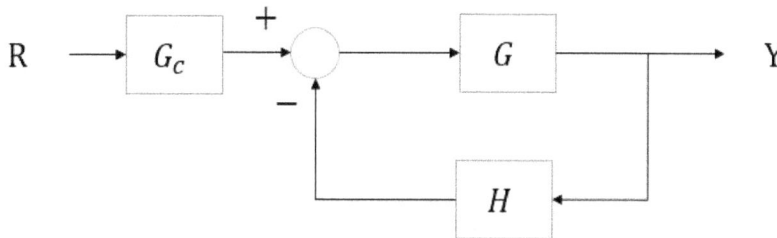

However, the __cascade type__ is commonly used in a compensator's design.

Let $G_c(s) = \dfrac{K(s + z)}{(s + p)}$ for the lead compensator:

$$G_c(j\omega) = \frac{K(j\omega + z)}{(j\omega + p)}$$

$$= \frac{K\left[1 + j\dfrac{\omega}{z}\right](z)}{\left[1 + j\dfrac{\omega}{p}\right](p)}$$

Let $\dfrac{1}{p} = \tau$, an $\dfrac{1}{z} = \alpha\tau$

$$\frac{z}{p} = \frac{1}{\alpha}; \quad K\left(\frac{z}{p}\right) = \frac{K}{\alpha}$$

Let $K_1 = \dfrac{K}{\alpha}$

So, $G_C(j\omega) = \dfrac{K_1(1 + j\omega\alpha\tau)}{(1 + j\omega\tau)}$

$$G_C(s) = \frac{K_1(1 + \alpha\tau s)}{(1 + \tau s)}$$

9.3.1. Lead Compensator Design

DESIGN PARAMETERS (α, τ)
Let $K_1 = 1$.

$$G_c(s) = \frac{1 + (\alpha\tau)s}{1 + (\tau)s} = \frac{\alpha\tau\left(s + \dfrac{1}{\alpha\tau}\right)}{\tau\left(s + \dfrac{1}{\tau}\right)} = \frac{\alpha\left(s + \dfrac{1}{\alpha\tau}\right)}{\left(s + \dfrac{1}{\tau}\right)}; \quad G_c(s) = \frac{\alpha(s + z)}{(s + p)}; \quad p > z$$

$$G_c(j\omega) = \frac{[1 + j\omega(\alpha\tau)]}{[1 + j\omega(\tau)]}$$

Magnitude: $M = 20\log\left|G_c(j\omega)\right| \, dB = 10\log\left\{\dfrac{(1 + \alpha^2\omega^2\tau^2)}{(1 + \omega^2\tau^2)}\right\} dB$

$$M = 10 \log C; \text{ where } C = \frac{1 + \alpha^2 \omega^2 \tau^2}{1 + \omega^2 \tau^2}$$

$$C = (10)^{(M/10)}$$

$$1 + \alpha^2 \omega^2 \tau^2 = C(1 + \omega^2 \tau^2)$$

$$\tau^2 \omega^2 (\alpha^2 - C) = C - 1$$

At $\omega = \omega_c, \boldsymbol{\tau = \dfrac{1}{\omega_c} \sqrt{\dfrac{C-1}{\alpha^2 - C}}}$

Phase: $\phi = \angle G_c = \tan^{-1}(\alpha \omega \tau) - \tan^{-1}(\omega \tau)$

In general: $\tan^{-1}(A) - \tan^{-1}(B) = \tan^{-1}\left\{ \dfrac{A-B}{1+AB} \right\}$

So, $\phi = \tan^{-1}\left\{ \dfrac{\omega\tau(\alpha - 1)}{1 + \alpha\omega^2\tau^2} \right\}$

$$\tan\phi = \frac{\omega\tau(\alpha - 1)}{1 + \alpha\omega^2\tau^2}$$

At $\omega = \omega_{max}$, $\tan\phi = \infty \Rightarrow 1 + \alpha\omega^2\tau^2 = 0; \left| \omega\tau \right| = \dfrac{1}{\sqrt{\alpha}}$

$$\tan\phi = \left(\frac{\alpha - 1}{\sqrt{\alpha}} \right)\left(\frac{1}{1+1} \right) = \frac{(\alpha - 1)}{2\sqrt{\alpha}}$$

$$\sin\phi = \frac{\tan\phi}{\sqrt{1 + \tan^2(\phi)}} = \frac{(\alpha - 1)}{2\sqrt{\alpha}} \frac{2\sqrt{\alpha}}{(\alpha + 1)} = \frac{\alpha - 1}{\alpha + 1}$$

So, we can find: $\alpha = \dfrac{1 + \sin\phi}{1 - \sin\phi}$

$$C = \frac{1 + \alpha^2 \omega^2 \tau^2}{1 + \omega^2 \tau^2}$$

Substituting: $\omega^2 \tau^2 = \dfrac{1}{\alpha}$

$$C = \frac{1+\alpha}{1+\left(\dfrac{1}{\alpha}\right)} = \alpha$$

So, $M = 10\log\alpha$; it is rounded off to the next highest integer.

Also, from $C = \dfrac{1+\alpha^2\omega^2\tau^2}{1+\omega^2\tau^2}$, we can write $\omega^2\tau^2 = \dfrac{1-C}{C-\alpha^2} = \dfrac{1}{\alpha}$ for $C = \alpha$.

Then, let $p = \tan\phi = \dfrac{\omega\tau(\alpha-1)}{1+\alpha\omega^2\tau^2}$, substituting for $\omega^2\tau^2 = \dfrac{1-C}{C-\alpha^2}$.

$$p^2 = \frac{(1-C)(C-\alpha^2)}{(\alpha+C)^2}$$

$$\alpha^2(p^2 - C + 1) + \alpha(2Cp^2) + (C^2p^2 + C^2 - C)$$

$$\alpha = \frac{1}{\left\{C-(p^2+1)\right\}}\left[Cp^2 \pm \sqrt{(Cp^2)^2 + \left\{C-(p^2+1)\right\}(C^2p^2 + C^2 - C)}\right]$$

Because it should be positive, $C > (p^2 + 1)$.

9.3.2. Design Process

Given the percent overshoot (PO) and settling time (T_s), we can find the values of ζ, and ω_n.

$$\zeta = \frac{\delta}{\sqrt{\delta^2 + \pi^2}}; \delta = \ln\left(\frac{100}{P.O.}\right)$$

$$\omega_n = \frac{4}{\zeta T_s}$$

$$\phi = 100\zeta = \text{ phase margin}$$

$$\alpha = \frac{1+\sin\phi}{1-\sin\phi}$$

$M = 10\log\alpha$ (round up to next highest integer)

$C = (10)^{(M/10)}$; it is not equal to α because M is rounded up to an integer.

$$\tau = \frac{1}{w_c}\sqrt{\frac{C-1}{\alpha^2 - C}}$$

w_c = gain crossover frequency and it either is given or can be found from a Bode plot.

Compensator design:

Compensator TF: $G_c = \dfrac{\alpha\left(s + \dfrac{1}{\alpha\tau}\right)}{\left(s + \dfrac{1}{\tau}\right)}$

Compensator zero: $s = -\dfrac{1}{\alpha\tau}$

Compensator pole: $s = -\dfrac{1}{\tau}$

9.3.3. Design Check

$$\phi > 0; \ M > 0; \ C > p^2 + 1; \ p = \tan\phi$$

The compensator pole should be greater than $|(-\zeta w_n)|$ and should be in the desired region.

If the compensator pole is not in the desired region, increase the gain crossover frequency (w_c) and repeat the process.

For a second-order system, the characteristic equation is $s^2 + (2\zeta w_n)s + w_n^2 = 0$.

For $\zeta < 1$, the roots (or poles) are $s_{1,2} = -\zeta w_n \pm jw_n\sqrt{1-\zeta^2} = -a \pm jb$.

$a = \zeta w_n; b = w_n\sqrt{1-\zeta^2}$

$$Cos\theta = \frac{|a|}{CA} = \frac{|a|}{\sqrt{a^2+b^2}}$$

$$CA = \sqrt{a^2+b^2} = w_n$$

$$Cos\theta = \frac{\zeta w_n}{w_n} = \zeta = \frac{|a|}{\sqrt{a^2+b^2}} = \frac{|\text{real part of pole}|}{\text{magnitude of pole}}$$

$\theta = \cos^{-1}\zeta$; For $a = 0$, $\zeta = 0$; For $b = 0$, $\zeta = 1$

The shaded area represents the desired region for the stability performance.

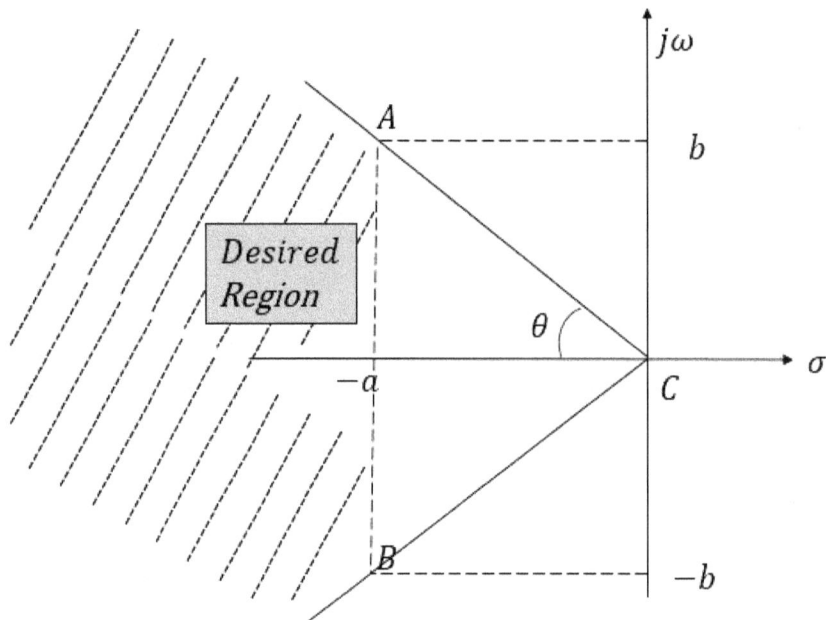

EXAMPLE

Design a compensator for given $\zeta = 0.45$, $T_s = 4$ sec and $w_c = 5$ rad/s

$$w_n = \frac{4}{\zeta T_s} = \frac{4}{0.45 \times 4} = 2.22$$

$$\phi = 100\zeta = 45°$$

$$\alpha = \frac{1 + \sin\phi}{1 - \sin\phi} = 5.83$$

$$M = 10\log\alpha = 10\log(5.83) = 7.66dB \approx 8dB \text{ (rounded up to the next highest value)}$$

$$C = (10)^{(M/10)} = (10)^{0.8} = 6.31$$

$$\tau = \frac{1}{w_c}\sqrt{\frac{C-1}{\alpha^2 - C}} = \frac{1}{5}\sqrt{\frac{5.31}{(5.83^2 - 6.31)}} = 0.087$$

$$G_c = \frac{\alpha\left(s + \dfrac{1}{\alpha\tau}\right)}{\left(s + \dfrac{1}{\tau}\right)} = \frac{5.83(s + 1.97)}{(s + 11.5)}$$

Zero: $s = -1.97$

Pole: $s = -11.5$

Check

$$\phi > 0,\ M > 0;\ p = \tan 45 = 1;\ p^2 + 1 = 2$$

$$C > (p^2 + 1); \zeta\omega_n = 1; \text{compensator pole}, \left|-p\right| > \left|(-\zeta\omega_n)\right|$$

$$\theta = \cos^{-1} 0.45 = 63.26°$$

The roots of the desired characteristic equation: $-\zeta\omega_n \pm j\omega_n\sqrt{1 - \zeta^2} = -1 \pm j1.98$

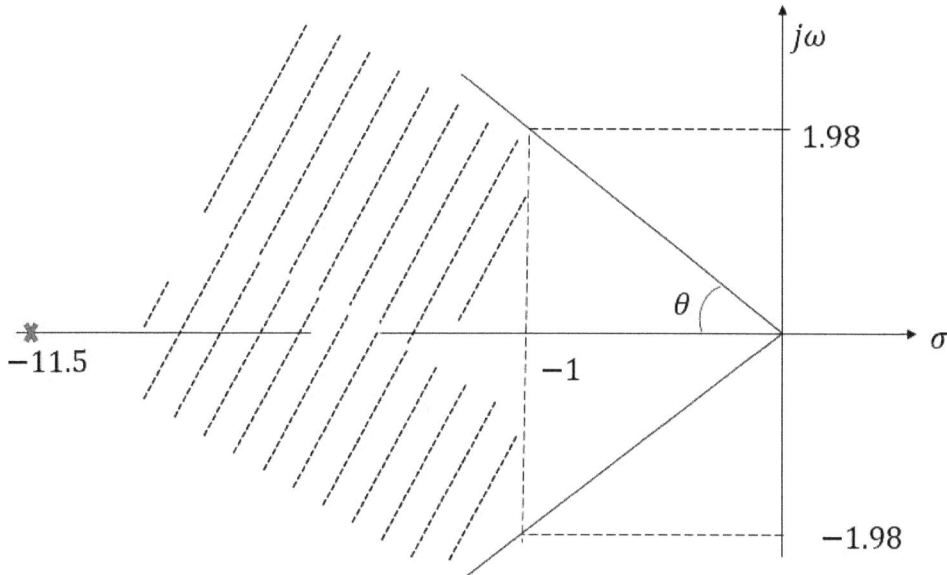

The pole is in the desired region.

EXAMPLE

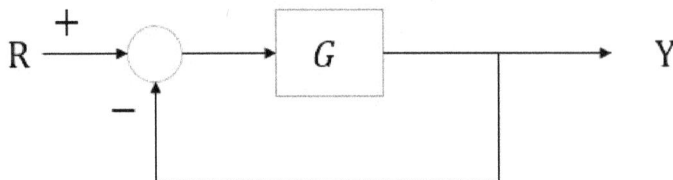

$$G(s) = \frac{10}{s(s+1)}$$

Desired value of $\zeta = 0.5$ and $T_s = 2.67$ sec

a) Check if a compensator is required.

b) If required, design a compensator:

(a) $T(s) = \dfrac{G}{1+G} = \dfrac{10}{s^2 + s + 10}$

Characteristic equation: $s^2 + s + 10 = 0$

Comparing with standard form: $s^2 + (2\zeta w_n)s + w_n^2 = 0$; $w_n = \sqrt{10} = 3.16$

$2\zeta w_n = 1$; $\zeta = \dfrac{1}{2w_n} = 0.16$; settling time, $T_s = \dfrac{4}{\zeta w_n} = 7.9$ sec

To increase the damping and reduce the settling time to the desired values, we need a compensator.

(b) Desired values are $\zeta = 0.5$, $T_s = 2.67$ sec.

$$w_n = \frac{4}{\zeta T_s} = \frac{4}{(0.5)(2.67)} = 3 \text{ rad/s}$$

Desired characteristic equation: $s^2 + (2\zeta w_n)s + w_n^2 = 0$

$$s^2 + 3s + 9 = 0$$

Desired poles: $s_{1,2} = -1.5 \pm j2.6$

$$\theta = \cos^{-1}\zeta = 60°; \ \zeta w_n = 1.5$$

CROSSOVER FREQUENCY (ω_c)

$$G(s) = \frac{10}{s^2 + s}$$

$$G(j\omega) = \frac{10}{-\omega^2 + j\omega}$$

$$\left|G(j\omega)\right| = \frac{10}{\sqrt{\omega^4 + \omega^2}} = 1$$

$$\omega^4 + \omega^2 - 100 = 0$$

$$\omega^2 = 0.5\left[-1 \pm \sqrt{1 + 400}\right]; \ \omega = 3.1, -10.5$$

$$\omega_c = 3.1$$

Compensator Design

$\zeta = 0.5, \ \omega_n = 3, \ \omega_c = 3.1$

$$G_c = \frac{\alpha\left(s + \dfrac{1}{\alpha\tau}\right)}{\left(s + \dfrac{1}{\tau}\right)}$$

$$\phi = 100\zeta = 50°$$

$$\alpha = \frac{1 + \sin\phi}{1 - \sin\phi} = 7.549$$

$$M = 10\log\alpha = 8.78 \approx 9 \ dB$$

$$C = (10)^{(M/10)} = 7.94$$

$$\tau = \frac{1}{\omega_c}\sqrt{\frac{C-1}{\alpha^2 - C}} = \frac{1}{3.1}\sqrt{\frac{6.94}{49.05}} = 0.12$$

$$G_c = 7.55\frac{(s + 1.1)}{(s + 8.33)}; \ \text{pole}, \ s = \left|-8.3\right| > \left|-1.5\right|$$

DESIRED REGION

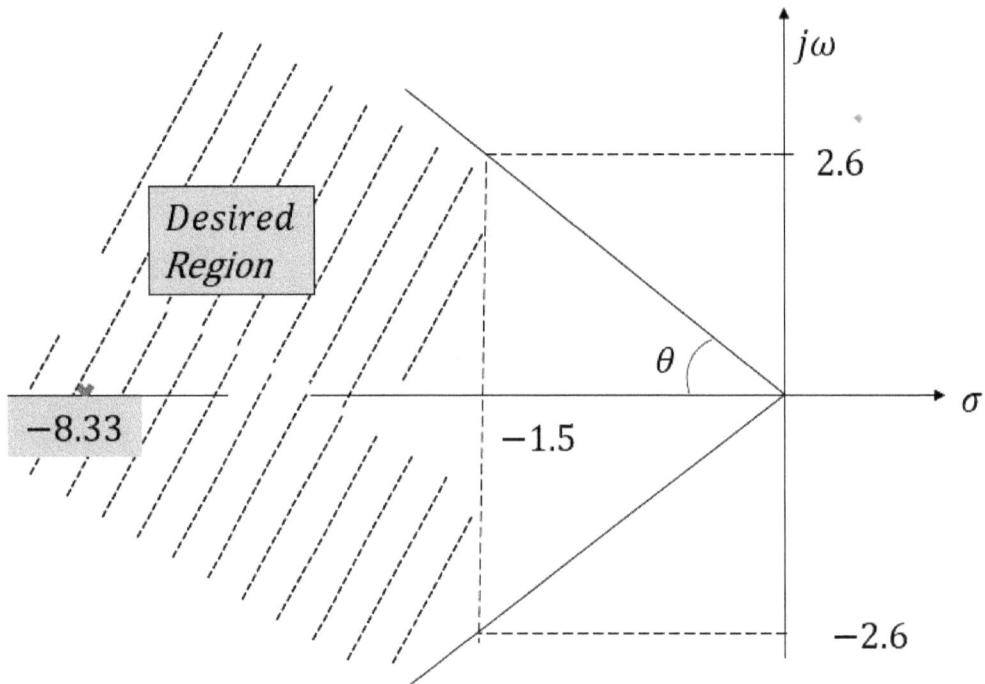

9.4. Prefilter Design

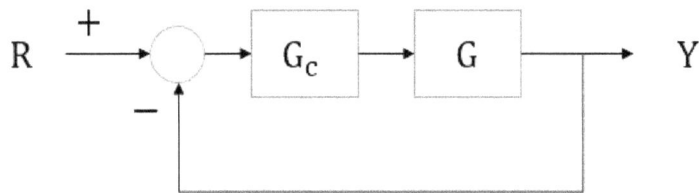

$$T(s) = \frac{G_c G}{1 + G_c G}; \text{ Let } G_c = \frac{s+z}{s+p} \text{ and } G = \frac{1}{s}$$

$$G_c G = \frac{(s+z)}{s(s+p)}; T(s) = \frac{(s+z)}{\{s(s+p)+(s+z)\}}$$

The zero of the system's TF, $T(s)$, is equal to the zero of the compensator's TF. Also, the zero of the compensator affects the system's characteristic equation and hence the system response. To eliminate this effect of zero in the compensator, a prefilter with a unit magnitude (gain) is required.

Prefilter With a P-I Compensator

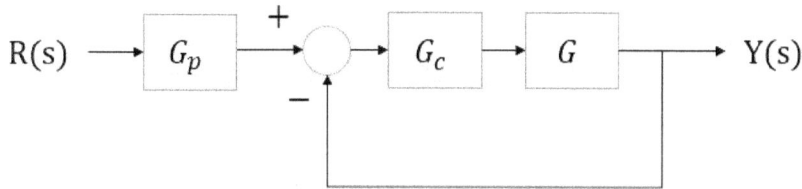

$$G_c(s) = K_P + \frac{K_I}{s} = \frac{K_P s + K_I}{s}; G(s) = \frac{1}{s}$$

$$T(s) = \frac{G_p G_c G}{(1 + G_c G)}; G_c G = \frac{K_P s + K_I}{s^2}$$

$$T(s) = G_p \left[\frac{K_P s + K_I}{s^2 + K_P s + K_I} \right] = G_p \left[\frac{K_P \left(s + \frac{K_I}{K_P} \right)}{(s^2 + K_P s + K_I)} \right]$$

To eliminate the zero, $\frac{K_I}{K_P}$, the TF of prefilter: $G_p = \frac{\left(\frac{K_I}{K_P} \right)}{\left(s + \frac{K_I}{K_P} \right)}$

The system's TF with a prefilter: $T(s) = \frac{K_I}{(s^2 + K_P s + K_I)}$

The system's TF without a prefilter: $T(s) = \left[\frac{K_P s + K_I}{s^2 + K_P s + K_I} \right]$

EXAMPLE
Design a prefilter for a P-I compensator if $PO = 4\%$, $T_s = 0.5$ sec.

$$\delta = \ln\left(\frac{100}{\text{P.O.}}\right) = \ln(25) = 3.22$$

$$\zeta = \frac{\delta}{\sqrt{\delta^2 + \pi^2}} = 0.716$$

$$\omega_n = \frac{4}{\zeta T_s} = 11.17 \text{ rad/sec}$$

Desired characteristic equation:

$$s^2 + (2\zeta\omega_n)s + \omega_n^2 = s^2 + 16s + 124.77$$

Comparing with $s^2 + K_p s + K_I$:

We can find $K_p = 16$, $K_I = 125$; $\dfrac{K_I}{K_P} = 7.8 \approx 8$.

So, the prefilter $G_p = \dfrac{8}{(s+8)}$.

Without prefilter:

$$T(s) = \frac{(K_p s + K_I)}{(s^2 + K_p s + K_I)} = \frac{(16s + 125)}{(s^2 + 16s + 125)}$$

With prefilter:

$$T(s) = \frac{125}{s^2 + 16s + 125}$$

So, in general, if $T(s)$ has a zero $(s + z)$ because of the addition of a compensator, the required prefilter to eliminate this zero is $G_p = \dfrac{z}{(s+z)}$.

EXAMPLE

Design a PI compensator and prefilter so that the control system satisfies the desired parameters of PO = 10% and $T_s = 2$ sec.

$$G(s) = \frac{1}{(s - 20)} \text{ and } H(s) = 10$$

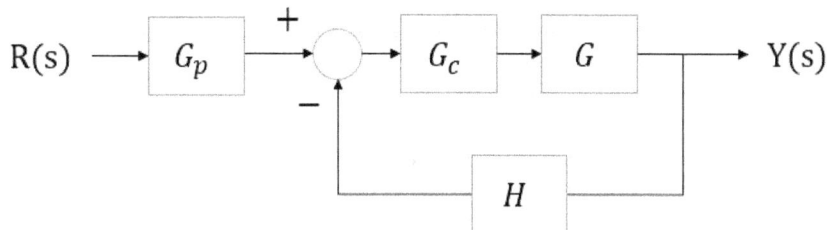

$$G_c(s) = K_P + \frac{K_I}{s} = \frac{(K_P s + K_I)}{s}$$

$$T(s) = \frac{G_p G_c G}{(1 + G_c G H)}; G_c G = \frac{(K_P s + K_I)}{s(s - 20)}$$

$$T(s) = \frac{G_p K_P \left(s + K_I/K_P\right)}{s(s - 20) + 10(K_P s + K_I)}$$

To eliminate the zero, $\left(s + K_I/K_P\right)$ the required prefilter should be $G_p = \frac{\left(K_I/K_P\right)}{\left\{s + \left(K_I/K_P\right)\right\}}$.

With prefilter:

$$T(s) = \frac{K_I}{s(s - 20) + 10(K_P s + K_I)}$$

Characteristic equation:

$$s(s - 20) + 10(K_P s + K_I) = 0$$

$$s^2 + (10K_P - 20)s + 10K_I = 0 \rightarrow (1)$$

Desired characteristic equation:

$$s^2 + (2\zeta\omega_n)s + \omega_n^2 = 0 \rightarrow (2)$$

The desired parameters are: PO = 10%, T_s = 2 sec

$$\delta = \ln\left(\frac{100}{P.O.}\right) = \ln 10 = 2.3; \zeta = \frac{\delta}{\sqrt{\delta^2 + \pi^2}} = 0.59$$

$$\omega_n = \frac{4}{\zeta T_s} = 3.39; \; 2\zeta\omega_n = 4$$

Comparing (1) and (2):

$$10K_P - 20 = 2\zeta\omega_n = 4; \; K_P = 2.4$$

$$10K_I = \omega_n^2 = 11.5; \; K_I = 1.15$$

$$\left(K_I/K_P\right) = 0.5; \; G_p = \frac{0.5}{s+0.5}; \; G_c = \left(\frac{1}{s}\right)(2.4 \; s + 1.15)$$

$$T(s) = \frac{1.15}{\left(s^2 + 4 \; s + 11.5\right)}$$

9.5. Lag Compensator Design

$$G_c(s) = K\left(\frac{s+z}{s+p}\right); \; p < z$$

$$\text{Let } \alpha = \frac{z}{p} \text{ and } \tau = \frac{1}{z}$$

$$p = \frac{z}{\alpha} = \frac{1}{\alpha\tau}$$

$$G_c = K\left[\frac{s+(1/\tau)}{s+(1/\alpha\tau)}\right] = K\alpha\frac{(1+\tau s)}{(1+\alpha\tau s)} = K_1\frac{(1+\tau s)}{(1+\alpha\tau s)}; \; K_1 = K\alpha$$

9.5.1. Design Process

For a given, G(s):

$$L(s) = G_c(s)G(s)$$

At $\omega = \omega_c$, magnitude in $\text{dB} = 20\log|L|$

$$z = \frac{\omega_c}{10}, \tau = \frac{1}{z}; \text{dB} = 20\log\alpha; \; \alpha = \text{inverse } \log(\text{dB}/20)$$

EXAMPLE

Given that $L(s) = G_c(s)G(s) = \dfrac{K_1}{s(s+2)}$:

Desired parameters: velocity error constant, $K_v = 20$ and crossover frequency, $\omega_c = 1.5$:

$$K_v = s\Big[L(s)\Big]_{s=0} = K_1/2;\ K_1 = 2K_v = 40$$

$$L(s) = \frac{40}{(s^2 + 2s)}$$

$$L\big(j\omega\big) = \frac{40}{(-\omega^2 + j2\omega)}$$

$$\text{At } \omega = \omega_c,\quad L(j\omega) = \frac{40}{(-2.25 + j3)}$$

$\text{dB} = 20\log\big|L(j\omega)\big|$

$= 20\log 40 - 10\log\{5.06 + 9\}$

$= 32 - 11.5 = 20.5\ dB$

$20\log\alpha = 20.5\ dB$

$\log\alpha = \dfrac{20.5}{20} \approx 1;\ \alpha = 10$

$$z = \frac{\omega_c}{10} = \frac{1.5}{10} = 0.15,\ \tau = \frac{1}{z} = 6.67,\ p = \frac{z}{\alpha} = 0.015$$

Compensator, $G_c = \dfrac{K_1(1+\tau s)}{(1+\alpha\tau s)} = \dfrac{40(1+6.67s)}{(1+66.67s)} = \dfrac{40\,(6.67)(s+0.15)}{(66.7)(s+0.015)} = \dfrac{4(s+0.15)}{(s+0.015)}$

Or, $G_c(s) = K\left(\dfrac{s+z}{s+p}\right)$, $\quad K = \dfrac{K_1}{\alpha} = 40/10 = 4;\ \ G_c(s) = \dfrac{4(s+0.15)}{(s+0.015)}$

EXAMPLE

Given that $L(s) = \dfrac{K_1}{s(s+10)^2}$:

Design a lag compensator for the desired parameters of $K_v = 20$, $w_c = 1.3$.

$$K_v = s\left[L(s)\right]_{s=0} = K_1/100; \ K_1 = 100 \ K_v = 2000$$

$$L(s) = \frac{2000}{(s^3 + 20 \ s^2 + 100s)}$$

$$L(j\omega) = \frac{2000}{-j\omega^3 - 20 \ \omega^2 + 100 j\omega}$$

At $w = w_c$, $L(j\omega) = \dfrac{2000}{-33.8 + j127.8}$:

$$dB = 20\log\left|L(j\omega)\right| = 20\log 2000 - 10\log\left\{(33.8)^2 + (127.8)^2\right\}$$

$$= 66 - 42 = 24$$

$$20\log\alpha = 24; \ \log\alpha = 1.2; \ \alpha = 16; \ z = \frac{w_c}{10} = 0.13; \ \tau = \frac{1}{z} = 7.69$$

$$p = \frac{z}{\alpha} = 0.01, \ \ K = \frac{K_1}{\alpha} = \frac{2000}{16} = 125$$

Compensator: $G_c = K\left(\dfrac{s+z}{s+p}\right) = 125\left\{\dfrac{s+0.13}{s+.01}\right\}$

Or $G_c = \dfrac{K_1(1+\tau s)}{(1+\alpha\tau s)} = \dfrac{2000(1+7.69s)}{(1+123s)} = \dfrac{2000 \ (7.69)(s+0.13)}{(123)(s+0.01)} = 125\left\{\dfrac{s+0.13}{s+.01}\right\}$

9.6. Summary

Cascade-type phase lead compensators are commonly used in control systems. However, the design process for both phase lead and phase lag compensators are shown in this chapter. A compensator can be designed for the given performance parameters of PO and settling time. The design's viability can be easily checked by verifying whether the compensator's pole is in the desired region for system stability. Adding a compensator generally affects the system's response by altering the characteristic equation of the system's TF. It is shown that it can be eliminated by adding a prefilter. So, the design of a prefilter with a P-I compensator is provided with examples in the study of control system design.

9.7. Assessment

1. The design parameter in the design of a compensator are:

    a. α

    b. τ

    c. α and τ

    d. None of the above

2. In a lead compensator, if p = pole and z = zero:

    a. $\alpha = p/z$

    b. $\alpha = z/p$

    c. $\alpha = pz$

    d. $\alpha = p + z$

3. In a lag compensator:

    a. $\alpha = p/z$

    b. $\alpha = z/p$

    c. $\alpha = pz$

    d. $\alpha = p + z$

4. In a lead compensator design:

    a. $\tau = zp$

    b. $\tau = 1/z$

    c. $\tau = 1/p$

    d. $\tau = pz$

5. In a lag compensator design:

    a. $\tau = 1/pz$

    b. $\tau = 1/z$

    c. $\tau = 1/p$

    d. $\tau = pz$

6. In a lead compensator design, the requirement for a pole is:

    a. $p > \zeta\omega_n$

    b. $p < \zeta\omega_n$

    c. $p > -\zeta\omega_n$

    d. $p < -\zeta\omega_n$

7. A prefilter is generally added to:

    a.   Eliminate a pole

    b.   Eliminate a zero

    c.   Add a zero

    d.   Add a pole

8. For a PI compensator, the prefilter is:

    a. $G_p = \dfrac{\left(K_I/K_p\right)}{\left(s + K_I/K_p\right)}$

    b. $G_p = \dfrac{\left(K_p/K_I\right)}{\left(s + K_p/K_I\right)}$

    c. $G_p = \dfrac{\left(K_p/K_I\right)}{\left(s + K_I/K_p\right)}$

    d. $G_p = \dfrac{\left(K_I/K_p\right)}{\left(s + K_p/K_I\right)}$

9. The prefilter for a system with $T(s) = \dfrac{s + 12}{s^2 + 7s + 12}$ is:

    a. $G_p = 12/(s + 12)$

    b. $G_p = 7/(s + 12)$

c. $G_p = 1/(s + 12)$

d. $G_p = 4/(s + 12)$

10. The velocity error constant for $L(s)$ is:

   a. $K_v = \left[L(s) \right]_{s = 0}$

   b. $K_v = s\left[L(s) \right]_{s = 0}$

   c. $K_v = s^2\left[L(s) \right]_{s = 0}$

   d. $K_v = \dfrac{1}{s}\left[L(s) \right]$

9.8. Practice Problems

1. Answer all the assessment questions in Section 9.7

2. Design a compensator for the control system given below such that the Percent Overshoot is 2% and settling time is 0.9 sec.

$$G(s) = \frac{25}{s(s + 2)}$$

Check your design whether the pole of compensator is in the desired region.

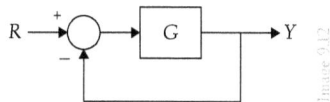

3. Design a compensator for a control system such that the damping ratio is 0.6, settling time is 1.5 sec., and the gain cross over frequency is 2.5 rad/sec. Also check your design whether the pole of compensator is in the desired region.

4. The desired damping ratio is 0.8 and settling time is 0.5 sec., for the control system given below.

$$G(s) = \frac{36}{s(s + 6)}$$

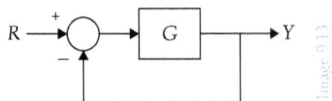

a. Check if a compensator is required and design one if required.

b. Check your design whether the pole of compensator is in the desired region.

5. Design a PI compensator such that the damping ratio is 0.25 and settling time is 2 sec, for the control system given below.

$$G(s) = \frac{1}{(s-10)}$$

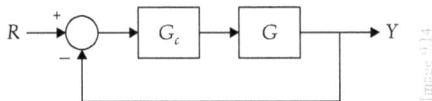

6. Design a PI compensator and a prefilter such that the percent overshoot is 5%, and settling time is 1 sec, for the control system given below.

$$G(s) = \frac{1}{(s-20)}$$

Determine the system transfer function with prefilter.

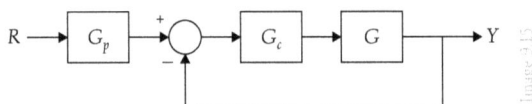

7. Design a prefilter for the control system with a compensator given below.

$$Compensator, G_c(s) = \frac{(4s+20)}{(2s+6)}; \quad G(s) = \frac{1}{s}$$

Determine the system transfer function with prefilter.

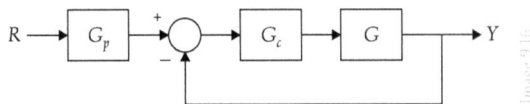

8. For a control system with a PI compensator having $K_P = 10$, and $K_1 = 80$, design a prefilter. The control system is given below.

$$G(s) = \frac{1}{(s-4)}$$

Determine the system transfer function with prefilter.

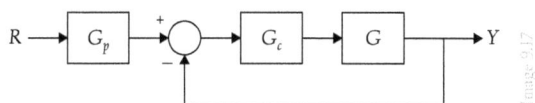

9. Design a Lag compensator for a control system given below, such that $K_v = 10$, $\omega_c = 1$

$$L(s) = \frac{K_1}{s(s+5)}$$

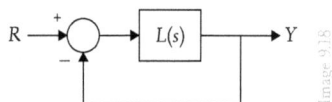

10. Design a Lag compensator for a control system given below, such that $K_v = 15$, $\omega_c = 1.5$

$$L(s) = \frac{K_1}{s(s+2)(s+6)}$$

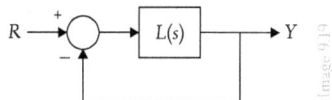

STATE VARIABLE MODELS AND DESIGN

10.1. Introduction

The state variable model is an alternate method to study control systems and helps reduce an nth-order system described by an nth-order differential equation into n number of first-order systems. The resulting equations of motion are called the *state equation* and *output equation*. They can be expressed in matrix form and solved with the help of a computer. Because this method can be applied to a large order of systems and solved by modern computers, it is also called a *modern control system*. This chapter discusses how to develop the state and output equations from either a differential equation or the transfer function (TF). Also, the block diagram or signal flow graph models can be easily developed for the state variable approach. The stability and design of control systems are included in the study of this system.

10.2. Learning Objectives

1. Model the nth-order system into n first-order systems using state variables.

2. Develop the state and output equations from a given differential equation of an nth-order system or TF.

3. Determine the block diagram or signal flow graph from given state and output equations.

4. Check the system's stability by the Routh-Hurwitz (R-H) criterion method.

5. Check the system for controllability and observability.

6. Design an observer and a controller for the given desired poles.

10.3. State Variable Model and nth-Order Systems

The nth-order control system is generally described by a single nth-order differential equation. The state variables help model the system into first-order systems described by an n number of first-order differential equations.

For example, a third-order system is described by:

$$\dddot{y} + a_2\ddot{y} + a_1\dot{y} + a_0 y = u(t), \text{ where } u(t) = \text{ step input forcing function}$$

$$\dddot{y} = -a_0 y - a_1\dot{y} - a_2\ddot{y} + u = -a_0 x_1 - a_1 x_2 - a_2 x_3 + u$$

where $x_1, x_2,$ and x_3 are defined as state variables.

$$x_1 = y; \; x_2 = \dot{y} = \dot{x}_1; x_3 = \ddot{y} = \dot{x}_2; \; \dot{x}_3 = \dddot{y}$$

We have three state variables because it is a third-order system. An nth-order system would require n number of state variables.

The state equations for this system are:

$$\dot{x}_1 = x_2; \; \dot{x}_2 = x_3; \; \dot{x}_3 = -a_0 x_1 - a_1 x_2 - a_2 x_3 + u$$

In matrix form:

$$\left\{ \begin{matrix} \dot{x}_1 \\ \dot{x}_2 \\ \dot{x}_3 \end{matrix} \right\} = \begin{bmatrix} 0 & 1 & 0 \\ 0 & 0 & 1 \\ -a_0 & -a_1 & -a_2 \end{bmatrix} \left\{ \begin{matrix} x_1 \\ x_2 \\ x_3 \end{matrix} \right\} + \left\{ \begin{matrix} 0 \\ 0 \\ 1 \end{matrix} \right\} u$$

$\{\dot{x}\} = [A]\{x\} + \{B\}u$ is the state equation

In this system, $x_1 = y$ relates the output y to the state variable x_1.

In matrix form:

$$y = \begin{bmatrix} 1 & 0 & 0 \end{bmatrix} \left\{ \begin{matrix} x_1 \\ x_2 \\ x_3 \end{matrix} \right\}$$

$$y = [C]\{x\}$$

The state equation and output equation represent a state variable model.

Solution to the State Equation

$$\{\dot{x}\} = [A]\{x\} + \{B\}u$$

Assuming that $\{x(0)\}$ is the given initial condition, the Laplace transform (LT) of the state equation gives:

$$\left[S\{X(s)\} - \{X(0)\} \right] = [A]\{X(s)\} + \{B\}U(s)$$

$$(S[I] - [A])\{X(s)\} = \{X(0)\} + \{B\}U(s)$$

$$[Z(s)]\{X(s)\} = \{X(0)\} + \{B\}U(s); [Z(s)] = S[I] - [A]$$

$$\{X(s)\} = [Z(s)]^{-1}\Big[\{X(0)\} + \{B\}U(s)\Big]$$

$$= [\phi(s)]\Big[\{X(0)\} + \{B\}U(s)\Big] = [\phi(s)]\{X(0)\} + \{\psi(s)\}$$

$$\text{where, } [\phi(s)] = [Z(s)]^{-1} = (S[I] - [A])^{-1}$$

$$\psi(s) = [\phi(s)]\{B\}U(s)$$

Time response: $\{X(s)\} = [\phi(s)]\{X(0)\} + \{\psi(s)\}$; taking the inverse LT

$\{X(t)\} = [\phi(t)]\{X(0)\} + \{\psi(t)\}$; $[\phi(t)] =$ state transition matrix or fundamental matrix

SPECIAL CASE

For $u(t) = 0$; $\psi(t) = 0$. Then, $\{X(t)\} = [\phi(t)]\{X(t)\}$

10.4. Time Response

Let the state equation be:

$$\left\{ \begin{array}{c} \dot{x}_1 \\ \dot{x}_2 \end{array} \right\} = \left[\begin{array}{cc} 0 & -2 \\ 1 & -3 \end{array} \right] \left\{ \begin{array}{c} x_1 \\ x_2 \end{array} \right\} + \left\{ \begin{array}{c} 0 \\ 1 \end{array} \right\} u(t)$$

Initial conditions: $\left\{ \begin{array}{c} x_1(0) \\ x_2(0) \end{array} \right\} = \left\{ \begin{array}{c} 1 \\ 1 \end{array} \right\}$

$$\{\dot{x}\} = [A]\{x\} + \{B\}u$$

$$\Big[Z(s)\Big] = S[I] - [A] = \left[\begin{array}{cc} s & 0 \\ 0 & s \end{array} \right] - \left[\begin{array}{cc} 0 & -2 \\ 1 & -3 \end{array} \right] = \left[\begin{array}{cc} s & 2 \\ -1 & s+3 \end{array} \right]$$

$$\Big[\phi(s)\Big] = \Big[Z(s)\Big]^{-1} = \frac{1}{\text{Det}} \left[\begin{array}{cc} s+3 & -2 \\ 1 & s \end{array} \right]$$

$$\text{Det} = s(s+3) + 2 = s^2 + 3s + 2 = (s+1)(s+2)$$

$$[\phi(s)] = \begin{bmatrix} \phi_{11} & \phi_{12} \\ \phi_{21} & \phi_{22} \end{bmatrix} = \begin{bmatrix} \dfrac{s+3}{(s+1)(s+2)} & \dfrac{-2}{(s+1)(s+2)} \\ \dfrac{1}{(s+1)(s+2)} & \dfrac{s}{(s+1)(s+2)} \end{bmatrix}$$

Taking the inverse LT for each element of the above matrix:

$$\phi_{11}(s) = \frac{s+3}{(s+1)(s+2)} = \frac{k_1}{s+1} + \frac{k_2}{s+2}$$

$$k_1 = (s+1)\phi_{11}(s)\,|_{s=-1} = 2; k_2 = (s+2)\phi_{11}(s)\,|_{s=-2} = -1$$

$$\phi_{11}(s) = \frac{2}{s+1} - \frac{1}{s+2}$$

$$\phi_{11}(t) = 2e^{-t} - e^{-2t}$$

$$\phi_{12}(s) = \frac{-2}{(s+1)(s+2)} = \frac{k_1}{s+1} + \frac{k_2}{s+2}$$

$$k_1 = (s+1)\phi_{12}(s)\,|_{s=-1} = -2; k_2 = (s+2)\phi_{12}(s)\,|_{s=-2} = 2$$

$$\phi_{12}(s) = -\frac{2}{s+1} + \frac{2}{s+2}$$

$$\phi_{12}(t) = -2e^{-t} + 2e^{-2t}$$

$$\phi_{21}(s) = \frac{1}{(s+1)(s+2)} = \frac{1}{s+1} - \frac{1}{s+2}$$

$$\phi_{21}(t) = e^{-t} - e^{-2t}$$

$$\phi_{22}(s) = \frac{s}{(s+1)(s+2)} = -\frac{1}{s+1} + \frac{2}{s+2}$$

$$\phi_{22}(t) = -e^{-t} + 2e^{-2t}$$

$$\psi(s) = [\phi(s)]\{B\}U(s);$$

For the unit step: $u(t) = 1$ and $U(s) = 1/s$

$$\{B\}U(s) = \left\{ \begin{array}{c} 0 \\ 1 \end{array} \right\}\{1/s\} = \left\{ \begin{array}{c} 0 \\ 1/s \end{array} \right\}$$

$$\left\{ \begin{array}{c} \psi_1(s) \\ \psi_2(s) \end{array} \right\} = \left[\begin{array}{cc} \phi_{11} & \phi_{12} \\ \phi_{21} & \phi_{22} \end{array} \right]\left[\begin{array}{c} 0 \\ 1/s \end{array} \right]$$

$$\psi_1(s) = \frac{1}{s}(\phi_{12}) = \frac{-2}{s(s+1)(s+2)} = -\frac{1}{s} + \frac{2}{s+1} - \frac{1}{s+2}$$

$$\psi_1(t) = -1 + 2e^{-t} - e^{-2t}$$

$$\psi_2(s) = \frac{1}{s}(\phi_{22}) = \frac{1}{s}\left[\frac{s}{(s+1)(s+2)} \right] = \frac{1}{(s+1)(s+2)} = \frac{1}{s+1} - \frac{1}{s+2}$$

$$\psi_2(t) = e^{-t} - e^{-2t}$$

$$\left\{ \begin{array}{c} x_1(t) \\ x_2(t) \end{array} \right\} = [\phi(t)]\left\{ \begin{array}{c} x_1(0) \\ x_2(0) \end{array} \right\} + \{\psi(t)\} = \left[\begin{array}{cc} \phi_{11} & \phi_{12} \\ \phi_{21} & \phi_{22} \end{array} \right]\left[\begin{array}{c} 1 \\ 1 \end{array} \right] + \left\{ \begin{array}{c} \psi_1(t) \\ \psi_2(t) \end{array} \right\}$$

$$x_1(t) = \phi_{11}(t) + \phi_{12}(t) + \psi_1(t) = 2e^{-t} - 1$$

$$x_2(t) = \phi_{21}(t) + \phi_{22}(t) + \psi_2(t) = e^{-t}$$

10.5. State Variable Model and Transfer Function

The state variable model can be developed from a given TF of a system.

In general, $T(s) = \dfrac{Y(s)}{R(s)}$; Let $R(s) = U(s)$ (unit step input).

Both $Y(s)$ and $R(s)$ are polynomials in s:

$$T(s) = \frac{Y(s)}{U(s)} = \frac{b_m S^m + b_{m-1}S^{m-1} + \cdots + b_1 S + b_0}{S^n + a_{n-1}S^{n-1} + \cdots + a_1 S + a_0} = \frac{\text{Output}}{\text{Input}}$$

$$= \frac{b_m S^m + b_{m-1} S^{m-1} + \cdots + b_1 S + b_0}{S^n \left[1 + a_{n-1} S^{-1} + \cdots + a_1 S^{-(n-1)} + a_0 S^{-n} \right]}$$

$$= \frac{b_m S^{-(n-m)} + b_{m-1} S^{-(n-m+1)} + \cdots + b_1 S^{-(n-1)} + b_0 S^{-n}}{1 + a_{n-1} S^{-1} + \cdots + a_1 S^{-(n-1)} + a_0 S^{-n}}$$

EXAMPLE

Determine the state and output equations for the TF given below.

$$T(s) = \frac{(2s^2 + 8s + 6)}{(s^3 + 8s^2 + 16s + 6)} = \frac{Y(s)}{U(s)}$$

SOLUTION

The denominator is a third-order polynomial, and hence it is a third-order system. So, it will have three state variables, x_1, x_2 and x_3:

$$T(s) = \frac{(2s^2 + 8s + 6)}{s^3 [1 + 8s^{-1} + 16s^{-2} + 6s^{-3}]}$$

$$= \frac{2s^{-1} + 8s^{-2} + 6s^{-3}}{1 + 8s^{-1} + 16s^{-2} + 6s^{-3}} = \frac{b_2 s^{-1} + b_1 s^{-2} + b_0 s^{-3}}{1 + a_2 s^{-1} + a_1 s^{-2} + a_0 s^{-3}}$$

State equations:

The state variables are: $x_1 = y$; $x_2 = \dot{y} = \dot{x}_1$; $x_3 = \ddot{y} = \dot{x}_2$.

The state equations are:

$$\dot{x}_1 = x_2; \; \dot{x}_2 = x_3; \; \dot{x}_3 = \dddot{y} = -a_0 x_1 - a_1 x_2 - a_2 x_3 + u = -6x_1 - 16x_2 - 8x_3 + u$$

In matrix form:

$$\begin{Bmatrix} \dot{x}_1 \\ \dot{x}_2 \\ \dot{x}_3 \end{Bmatrix} = \begin{bmatrix} 0 & 1 & 0 \\ 0 & 0 & 1 \\ -6 & -16 & -8 \end{bmatrix} \begin{Bmatrix} x_1 \\ x_2 \\ x_3 \end{Bmatrix} + \begin{Bmatrix} 0 \\ 0 \\ 1 \end{Bmatrix} u$$

Output equation:

$$y(t) = [b_0 \quad b_1 \quad b_2] \begin{Bmatrix} x_1 \\ x_2 \\ x_3 \end{Bmatrix} = [6 \quad 8 \quad 2] \begin{Bmatrix} x_1 \\ x_2 \\ x_3 \end{Bmatrix}$$

The numerator polynomial gives the output equation, while the denominator polynomial provides the state equation.

10.6. Block Diagram and Signal Flow Graph

Taking the LT of the state variables in the above example:

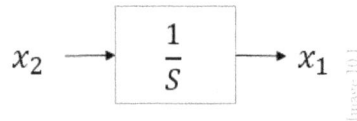

$$\dot{x}_1 = x_2; \; sx_1 = x_2; \; x_1 = x_2/s$$

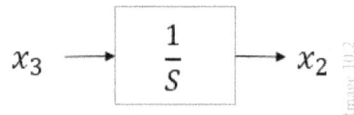

$$\dot{x}_2 = x_3; \; sx_2 = x_3; \; x_2 = x_3/s$$

$$\dot{x}_3 = -6x_1 - 16x_2 - 8x_3 + u; \; sx_3 = -6x_1 - 16x_2 - 8x_3 + U$$

$$x_3 = \frac{1}{s}(-6x_1 - 16x_2 - 8x_3 + U)$$

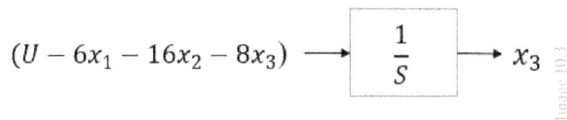

Output equation: $Y(t) = 6x_1 + 8x_2 + 2x_3;$ $Y(s) = 6x_1 + 8x_2 + 2x_3$

Block Diagram

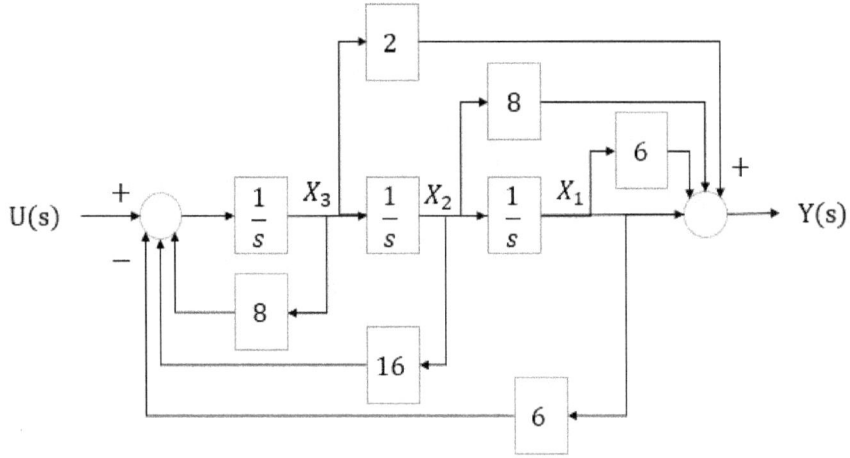

Signal Flow Graph

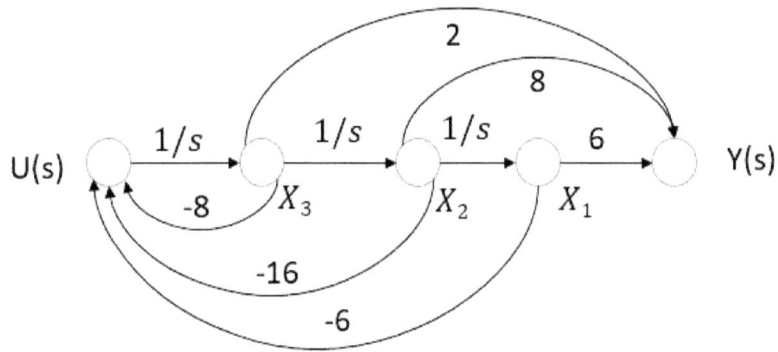

EXAMPLE

$$T(s) = \frac{4s + 12}{s^2 + 8s + 12} = \frac{4s^{-1} + 12s^{-2}}{1 + 8s^{-1} + 12s^{-2}}$$

Standard format:

$$T(s) = \frac{b_1 s^{-1} + b_0 s^{-2}}{1 + a_1 s^{-1} + a_0 s^{-2}}$$

The denominator is a second-order polynomial, and hence it requires two state variables, x_1 and x_2.

State equation:

$$\left\{ \begin{array}{c} \dot{x}_1 \\ \dot{x}_2 \end{array} \right\} = \left[\begin{array}{cc} 0 & 1 \\ -a_0 & -a_1 \end{array} \right] \left\{ \begin{array}{c} x_1 \\ x_2 \end{array} \right\} + \left\{ \begin{array}{c} 0 \\ 1 \end{array} \right\} u = \left[\begin{array}{cc} 0 & 1 \\ -12 & -8 \end{array} \right] \left\{ \begin{array}{c} x_1 \\ x_2 \end{array} \right\} + \left\{ \begin{array}{c} 0 \\ 1 \end{array} \right\} u$$

Output equation:

$$y(t) = [\begin{array}{cc} b_0 & b_1 \end{array}] \left\{ \begin{array}{c} x_1 \\ x_2 \end{array} \right\} = [\begin{array}{cc} 12 & 4 \end{array}] \left\{ \begin{array}{c} x_1 \\ x_2 \end{array} \right\} = 12x_1 + 4x_2$$

Block Diagram

$$\dot{x}_1 = x_2; \quad sx_1 = x_2 \qquad\qquad x_2 \longrightarrow \boxed{\dfrac{1}{s}} \longrightarrow x_1$$

$$\dot{x}_2 = -12x_1 - 8x_2 + u; \quad sx_2 = (U - 12x_1 - 8x_2)$$

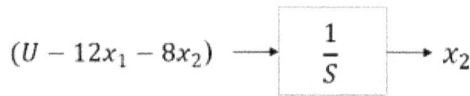

$$(U - 12x_1 - 8x_2) \longrightarrow \boxed{\dfrac{1}{s}} \longrightarrow x_2$$

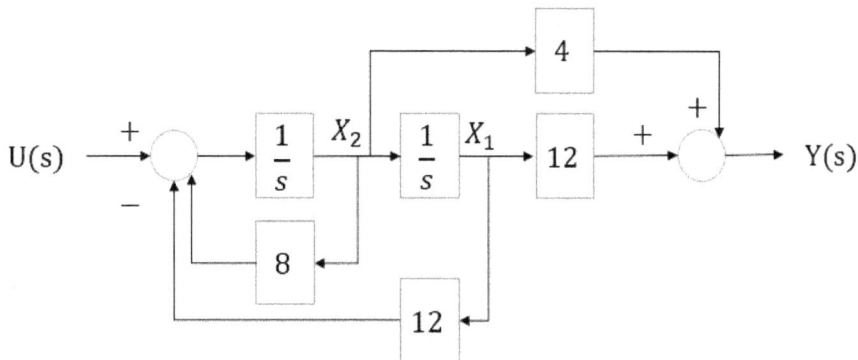

Signal Flow Graph

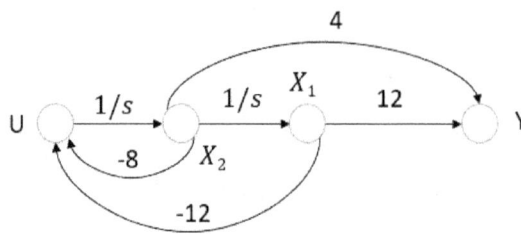

$$x_2 = \frac{1}{s}(U - 8x_2 - 12x_1)$$

$$x_1 = \frac{1}{s}(x_2)$$

$$Y = 12x_1 + 4x_2$$

EXAMPLE

$$\begin{bmatrix} \dot{x}_1 \\ \dot{x}_2 \end{bmatrix} = \begin{bmatrix} 0 & 1 \\ (2+k_1) & (3-k_2) \end{bmatrix} \begin{bmatrix} x_1 \\ x_2 \end{bmatrix} + \begin{bmatrix} 0 \\ 1 \end{bmatrix} u(t)$$

Determine k_1 and k_2 so that the desired roots of characteristic equation are -2 and -3.

Characteristic equation: $\left|z(s)\right| = 0$

$$\left[z(s)\right] = S[I] - [A] = \begin{bmatrix} s & 0 \\ 0 & s \end{bmatrix} - \begin{bmatrix} 0 & 1 \\ (2+k_1) & (3-k_2) \end{bmatrix} = \begin{bmatrix} s & -1 \\ -(2+k_1) & (s-3+k_2) \end{bmatrix}$$

$$\left|z(s)\right| = s(s-3+k_2) - (2+k_1) = 0$$

$$s^2 - 3s + k_2 s - 2 - k_1 = 0$$

$$s^2 + (k_2 - 3)s - (2+k_1) = 0 \rightarrow (i)$$

For given roots, the characteristic equation is:

$$(s - p_1)(s - p_2) = 0; p_1 = -2, p_2 = -3$$

$$(s+2)(s+3) = s^2 + 5s + 6 = 0 \rightarrow (ii)$$

Comparing (i) and (ii):

$$k_2 - 3 = 5 \Rightarrow k_2 = 8; \qquad -2 - k_1 = 6 \Rightarrow k_1 = -8$$

Check for the desired roots (s = −2, −3):

$$\dot{x}_1 = x_2$$

$$\dot{x}_2 = (2 + k_1) x_1 + (3 - k_2) x_2 + u$$

$$\ddot{x}_1 = -6 x_1 - 5 \dot{x}_1 + u$$

$$u = \ddot{x}_1 + 5 \dot{x}_1 + 6 x_1 ; U(s) = (s^2 + 5s + 6) X_1(s)$$

$$y = x_1; Y(s) = X_1(s)$$

TF: $T(s) = \dfrac{Y(s)}{U(s)} = \dfrac{1}{(s^2 + 5s + 6)} = \dfrac{1}{(s + 2)(s + 3)}$

Characteristic equation: $(s + 2)(s + 3) = 0$

The roots are $s = -2, -3$.

10.7. Stability of a State Variable Model

State equation:

$$\{\dot{x}\} = [A]\{x\} + \{B\}u$$

$$[Z(s)] = S[I] - [A]$$

Characteristic equation: $q(s) = |Z(s)| = 0$

Now, the R-H criterion can be applied to the characteristic equation for a stability analysis.

EXAMPLE

$$\begin{Bmatrix} \dot{x}_1 \\ \dot{x}_2 \\ \dot{x}_3 \end{Bmatrix} = \begin{bmatrix} 0 & 1 & 0 \\ 0 & 0 & 1 \\ -6 & -16 & -8 \end{bmatrix} \begin{Bmatrix} x_1 \\ x_2 \\ x_3 \end{Bmatrix} + \begin{Bmatrix} 0 \\ 0 \\ 1 \end{Bmatrix} u$$

$$\{\dot{x}\} = [A]\{x\} + \{B\}u$$

$$[Z(s)] = S[I] - [A] = \begin{bmatrix} s & -1 & 0 \\ 0 & s & -1 \\ 6 & 16 & s+8 \end{bmatrix}$$

Characteristic equation $= \left| Z(s) \right| = 0$

$$\left| Z(s) \right| = s^3 + 8s^2 + 16s + 6 = 0$$

Routh Array

s^3	1	16
s^2	8	6
s^1	15.25	0
s^0	6	

The system is stable because the first column of the array is positive.

EXAMPLE

Determine k for the system to be stable:

$$\begin{Bmatrix} \dot{x}_1 \\ \dot{x}_2 \end{Bmatrix} = \begin{bmatrix} -3 & 1 \\ -k & 1 \end{bmatrix} \begin{Bmatrix} x_1 \\ x_2 \end{Bmatrix} + \begin{Bmatrix} 0 \\ 1 \end{Bmatrix} u$$

$$[Z(s)] = S[I] - [A] = \begin{bmatrix} s+3 & -1 \\ k & s-1 \end{bmatrix}$$

Characteristic equation: $\left| Z(s) \right| = (s+3)(s-1) + k = s^2 + 2s + (k-3)$

Routh Array

s^2	1	$k-3$
s^1	2	0
s^0	$k-3$	

For the first column to be positive, $k > 3$. So, the value of k should be greater than three for the system to be stable.

10.8. Design of a State Variable System

Controllability and Observability

Controllability is the ability of a control system to reach a definite state from an initial state in a finite time. It defines the behavior of the control system. A system is completely controllable if the determinant of the controllability matrix is not zero.

Observability is the ability of a control system to determine its internal states by observing the output in a finite time interval for a given input. It shows the control system's behavioral approach. A system is completely observable if the determinant of the observability matrix is not zero.

Controllability:

The state equation of a system is given as:

$$\{\dot{x}\} = [A]\{x\} + \{B\}u$$

The controllability matrix: $[p_c] = [\{p_1\}\{p_2\} \cdots \{p_N\}]$

Where $\{p_1\} = \{B\}$; $\{p_2\} = [A]\{p_1\}$; $\{p_3\} = [A]\{p_2\}$

$$\{p_N\} = [A]\{p_{N-1}\}$$

If $|p_c| = 0$, the system is not controllable, and if $|p_c| \neq 0$, the system is controllable.

Observability:

The output equation of a system is: $y = [C]\{x\}$.

The observability matrix: $[p_o] = \begin{vmatrix} [p_1] \\ \vdots \\ [p_N] \end{vmatrix}$

Where, $[p_1] = [C]$; $[p_2] = [p_1][A]$; $[p_3] = [p_2][A]$

$$[p_N] = [p_{N-1}][A]$$

If $|p_o| = 0$, the system is not observable, and if $|p_o| \neq 0$, the system is observable.

EXAMPLE

$$\dddot{y} + 4\ddot{y} + 3\dot{y} + 2y = u$$

Check for controllability of the system:

$$\dddot{y} = u - 2y - 3\dot{y} - 4\ddot{y}; \text{ let } x_1 = y, x_2 = \dot{y}, x_3 = \ddot{y}$$

$$\dddot{y} = u - 2x_1 - 3x_2 - 4x_3$$

State equation: $\dot{x}_1 = x_2; \dot{x}_2 = x_3; \dot{x}_3 = \dddot{y} = u - 2x_1 - 3x_2 - 4x_3$

$$\begin{Bmatrix} \dot{x}_1 \\ \dot{x}_2 \\ \dot{x}_3 \end{Bmatrix} = \begin{bmatrix} 0 & 1 & 0 \\ 0 & 0 & 1 \\ -2 & -3 & -4 \end{bmatrix} \begin{bmatrix} x_1 \\ x_2 \\ x_3 \end{bmatrix} + \begin{Bmatrix} 0 \\ 0 \\ 1 \end{Bmatrix} u$$

$$[p_c] = [\{p_1\}\{p_2\}\{p_3\}]$$

$$\{p_1\} = \{B\} = \begin{Bmatrix} 0 \\ 0 \\ 1 \end{Bmatrix}$$

$$\{p_2\} = [A]\{p_1\} = \begin{bmatrix} 0 & 1 & 0 \\ 0 & 0 & 1 \\ -2 & -3 & -4 \end{bmatrix} \begin{Bmatrix} 0 \\ 0 \\ 1 \end{Bmatrix} = \begin{Bmatrix} 0 \\ 1 \\ -4 \end{Bmatrix}$$

$$\{p_3\} = [A]\{p_2\} = \begin{bmatrix} 0 & 1 & 0 \\ 0 & 0 & 1 \\ -2 & -3 & -4 \end{bmatrix} \begin{Bmatrix} 0 \\ 1 \\ -4 \end{Bmatrix} = \begin{Bmatrix} 1 \\ -4 \\ 13 \end{Bmatrix}$$

$$[p_c] = \begin{bmatrix} 0 & 0 & 1 \\ 0 & 1 & -4 \\ 1 & -4 & 13 \end{bmatrix}$$

$$\left| p_c \right| = 1(0 - 1) = -1$$

Because $\left| p_c \right| \neq 0$, the system is controllable.

EXAMPLE

$$\left\{ \begin{array}{c} \dot{x}_1 \\ \dot{x}_2 \end{array} \right\} = \left[\begin{array}{cc} -2 & 0 \\ d & -3 \end{array} \right] \left\{ \begin{array}{c} x_1 \\ x_2 \end{array} \right\} + \left\{ \begin{array}{c} 1 \\ 0 \end{array} \right\} u$$

Find d so that the system is controllable:

$$[p_c] = [\{p_1\}\{p_2\}]$$

$$\{p_1\} = \{B\} = \left\{ \begin{array}{c} 1 \\ 0 \end{array} \right\}$$

$$\{p_2\} = [A]\{p_1\} = \left[\begin{array}{cc} -2 & 0 \\ d & -3 \end{array} \right] \left\{ \begin{array}{c} 1 \\ 0 \end{array} \right\} = \left\{ \begin{array}{c} -2 \\ d \end{array} \right\}$$

$$[p_c] = \left[\begin{array}{cc} 1 & -2 \\ 0 & d \end{array} \right]$$

The determinant: $\left| p_c \right| = d$

For the system to be controllable, $\left| p_c \right| \neq 0$. So $d \neq 0$.

EXAMPLE

For a given output equation, $y = [C]\{x\}$, where $[C] = [1 \ \ 0 \ \ 0]$,

and a given state equation as:

$$\{\dot{x}\} = [A]\{x\} + \{B\}u$$

$$\text{where } [A] = \left[\begin{array}{ccc} 0 & 1 & 0 \\ 0 & 0 & 1 \\ -2 & -3 & -4 \end{array} \right] \text{ and } \{B\} = \left\{ \begin{array}{c} 0 \\ 0 \\ 1 \end{array} \right\}$$

Determine if the system is completely observable:

$$[p_o] = \left[\begin{array}{c} [p_1] \\ [p_2] \\ [p_3] \end{array} \right]$$

$$[p_1] = [C] = [1 \quad 0 \quad 0]$$

$$[p_2] = [p_1][A] = [1 \quad 0 \quad 0] \begin{bmatrix} 0 & 1 & 0 \\ 0 & 0 & 1 \\ -2 & -3 & -4 \end{bmatrix} = [0 \quad 1 \quad 0]$$

$$[p_3] = [p_2][A] = [0 \quad 1 \quad 0] \begin{bmatrix} 0 & 1 & 0 \\ 0 & 0 & 1 \\ -2 & -3 & -4 \end{bmatrix} = [0 \quad 0 \quad 1]$$

$$[p_o] = \begin{bmatrix} 1 & 0 & 0 \\ 0 & 1 & 0 \\ 0 & 0 & 1 \end{bmatrix}; |p_o| = 1$$

Because $p_o \neq 0$, the system is completely observable.

EXAMPLE

Determine the controllability and observability of a system with the given state and output equations:

$$\left\{ \begin{array}{c} \dot{x}_1 \\ \dot{x}_2 \end{array} \right\} = \begin{bmatrix} 2 & 0 \\ -1 & 1 \end{bmatrix} \left\{ \begin{array}{c} x_1 \\ x_2 \end{array} \right\} + \left\{ \begin{array}{c} 1 \\ -1 \end{array} \right\} u; \quad y = [1 \quad 0] \left\{ \begin{array}{c} x_1 \\ x_2 \end{array} \right\}$$

Controllability check:

$$[p_c] = [\{p_1\}\{p_2\}]$$

$$\{p_1\} = \{B\} = \left\{ \begin{array}{c} 1 \\ -1 \end{array} \right\}$$

$$\{p_2\} = [A]\{p_1\} = \begin{bmatrix} -2 & 0 \\ -1 & 1 \end{bmatrix} \left\{ \begin{array}{c} 1 \\ -1 \end{array} \right\} = \left\{ \begin{array}{c} 2 \\ -2 \end{array} \right\}$$

$$[p_c] = \begin{bmatrix} 1 & 2 \\ -1 & -2 \end{bmatrix}; |p_c| = 0;$$

So, the system is not controllable.

Observability check:

$$[p_o] = \begin{bmatrix} [p_1] \\ [p_2] \end{bmatrix}$$

$$[p_1] = [C] = [1 \quad 0]$$

$$[p_2] = [p_1][A] = [\ 1 \quad 0\ ] \begin{bmatrix} 2 & 0 \\ -1 & 1 \end{bmatrix} = [\ 2 \quad 0\ ]$$

$$[p_o] = \begin{bmatrix} 1 & 0 \\ 2 & 0 \end{bmatrix}; \ |p_o| = 0$$

So, the system is not observable.

10.9. Design of a Controller and an Observer

In state variable model control systems, the process for the design of a controller and an observer are given here. The design is based on the pole placement technique for an acceptable performance of a feedback system. The first step in the design process is to check if a system is completely controllable and observable.

Controller Design

The design objective is to find the controller gain matrix, [K], for a given pole location.

Then, the system input, $u(t) = -[K]\{x\}$,

where [K] is the controller gain, and it is a row matrix.

For an nth-order system, $[K] = [K_1\ K_2\ K_3 \dots\dots K_n]$

$$\{\dot{x}\} = [A]\{x\} + \{B\}u;\ u = -[K]\{x\}$$

$$\{\dot{x}\} = [[A] - \{B\}[K]]\{x\}$$

$$\{\dot{x}\} = [A_1]\{x\};\ [A_1] = [A] - \{B\}[K]$$

Let the solution be: $\{x\} = e^{st}$

$$\{\dot{x}\} = se^{st} = s\{x\}$$

$$s\{x\} = [A_1]\{x\}$$

$$\left\langle s[I] - [A_1] \right\rangle\{x\} = 0$$

The characteristic equation: determinant, $\left| sI - A_1 \right| = 0 \rightarrow (1)$

For a second-order system, if the given poles are s_1 and s_2, the desired characteristic equation is $(s - s_1)(s - s_2) = 0 \rightarrow (2)$.

The controller gain can be determined by comparing equations (1) and (2).

EXAMPLE

$$\left\{\begin{array}{c} \dot{x}_1 \\ \dot{x}_2 \end{array}\right\} = \left[\begin{array}{cc} 0 & 1 \\ 0 & 0 \end{array}\right]\left[\begin{array}{c} x_1 \\ x_2 \end{array}\right] + \left\{\begin{array}{c} 0 \\ 1 \end{array}\right\}u; \ y = \left[\begin{array}{cc} 1 & 0 \end{array}\right]\left\{\begin{array}{c} x_1 \\ x_2 \end{array}\right\}$$

Design a controller so that the desired poles are $s_1 = -1 + j$ and $s_2 = -1 - j$.

$$[A_1] = [A] - \{B\}[K]; \ [K] = [K_1 \ K_2]$$

$$[A_1] = \left[\begin{array}{cc} 0 & 1 \\ 0 & 0 \end{array}\right] - \left\{\begin{array}{c} 0 \\ 1 \end{array}\right\}\left[\begin{array}{cc} K_1 & K_2 \end{array}\right] = \left[\begin{array}{cc} 0 & 1 \\ 0 & 0 \end{array}\right] - \left[\begin{array}{cc} 0 & 0 \\ K_1 & K_2 \end{array}\right] = \left[\begin{array}{cc} 0 & 1 \\ -K_1 & -K_2 \end{array}\right]$$

$$s[I] - [A_1] = \left[\begin{array}{cc} s & 0 \\ 0 & s \end{array}\right] - \left[\begin{array}{cc} 0 & 1 \\ -K_1 & -K_2 \end{array}\right] = \left[\begin{array}{cc} s & -1 \\ K_1 & s + K_2 \end{array}\right]$$

$$\left| sI - A_1 \right| = 0; \ s^2 + sK_2 + K_1 = 0 \rightarrow (1)$$

For the given poles, $(s - s_1)(s - s_2) = s^2 - (s_1 + s_2)s + s_1 s_2 = 0 \rightarrow (2)$.

Comparing equations (1) and (2):

$$K_1 = s_1 s_2 = (-1 + j)(-1 - j) = 1 - (-1) = 2$$

$$K_2 = -(s_1 + s_2) = -(-1 + j - 1 - j) = 2$$

The controller gain matrix: $[K] = [\ 2\quad 2\ ]$

Comparing equation (2) with the standard characteristic equation of:

$$s^2 + 2\zeta\omega_n s + \omega_n^{\ 2} = 0$$

$$\omega_n = \sqrt{s_1 s_2} = \sqrt{2}$$

$$\zeta = -\frac{(s_1 + s_2)}{2\omega_n} = \frac{1}{\sqrt{2}}$$

EXAMPLE

Design a controller for a given third-order system of $\dddot{y} + 5\ddot{y} + 3\dot{y} + 2y = u$ so that the three

poles of the characteristic equation are, $s_1 = -4.8$, $s_{2,3} = -4.8 \pm j3.6$

$$\dddot{y} = u - 2y - 3\dot{y} - 5\ddot{y}$$

State equation:

$$\begin{Bmatrix} \dot{x}_1 \\ \dot{x}_2 \\ \dot{x}_3 \end{Bmatrix} = \begin{bmatrix} 0 & 1 & 0 \\ 0 & 0 & 1 \\ -2 & -3 & -5 \end{bmatrix} \begin{Bmatrix} x_1 \\ x_2 \\ x_3 \end{Bmatrix} + \begin{Bmatrix} 0 \\ 0 \\ 1 \end{Bmatrix} u$$

Output equation:

$$y = [1\quad 0\quad 0] \begin{Bmatrix} x_1 \\ x_2 \\ x_3 \end{Bmatrix}$$

Let the controller gain matrix be: $[K] = [K_1\quad K_2\quad K_3]$

$$\{B\}[K] = \begin{Bmatrix} 0 \\ 0 \\ 1 \end{Bmatrix} [K_1\quad K_2\quad K_3] = \begin{bmatrix} 0 & 0 & 0 \\ 0 & 0 & 0 \\ K_1 & K_2 & K_3 \end{bmatrix}$$

$$[A_1] = [A] - \{B\}[K] = \begin{bmatrix} 0 & 1 & 0 \\ 0 & 0 & 1 \\ -2 & -3 & -5 \end{bmatrix} - \begin{bmatrix} 0 & 0 & 0 \\ 0 & 0 & 0 \\ K_1 & K_2 & K_3 \end{bmatrix}$$

$$= \begin{bmatrix} 0 & 1 & 0 \\ 0 & 0 & 1 \\ -2 - K_1 & -3 - K_2 & -5 - K_3 \end{bmatrix}$$

$$|sI - A_1| = \begin{vmatrix} s & -1 & 0 \\ 0 & s & -1 \\ 2 + K_1 & 3 + K_2 & s + 5 + K_3 \end{vmatrix} = 0$$

$$s^3 + (5 + K_3)s^2 + (3 + K_2)s + (2 + K_1) = 0 \rightarrow (1)$$

The desired characteristic equation for the given poles:

$$(s - s_1)(s - s_2)(s - s_3) = 0$$

$$s^3 + 14.4s^2 + 82.08s + 172.8 = 0 \rightarrow (2)$$

Comparing equations (1) and (2):

$$K_1 + 2 = 172.8;\ K_1 = 170.8$$

$$K_2 + 3 = 82.08;\ K_2 = 79.08$$

$$K_3 + 5 = 14.4;\ K_3 = 9.4$$

Controller gain matrix: $[K] = [170.8 \quad 79.1 \quad 9.4]$

The desired characteristic equation for a given one real pole and a pair of complex poles:

$$(s + \zeta \omega_n)(s^2 + 2\zeta \omega_n s + \omega_n^2) = 0$$

$$s^3 + (3\zeta \omega_n)s^2 + \omega_n^2(1 + 2\zeta^2)s + (\zeta \omega_n^3) = 0$$

Comparing with equation (2):

$$3\zeta \omega_n = 14.4;\ \zeta \omega_n = 4.8$$

$$\zeta \omega_n^{\ 3} = 172.8 = \omega_n^{\ 2}\ (\zeta \omega_n);\ \omega_n^{\ 2} = 172.8/4.8 = 36;\ \omega_n = 6$$

$$\zeta = \frac{4.8}{6} = 0.8$$

$$\omega_n^{\ 2}(1 + 2\zeta^2) = 82.08$$

The percent overshoot (PO) and settling time (T_S) can be determined from the values of ω_n and ζ.

$$PO = 100e^{\frac{-\zeta \pi}{\sqrt{1-\zeta^2}}} = 1.52\%;\ T_s = \frac{4}{\zeta \omega_n} = 0.83\ \text{sec}$$

Ackermann's Formula Method

This method is viable for higher-order systems and can easily be programmed into a computer for determining the controller gain matrix, $[K]$.

For an nth-order system, $[K] = [K_1\quad K_2\quad \cdots\quad K_n]$.

Then, Ackermann's formula for the controller gain matrix is:

$$[K] = [0\quad 0\quad \cdots\quad 1][P_c]^{-1}[q(A)]$$

where $[P_c]$ is the controllability matrix.

If the desired poles are $s_1, s_2, \cdots s_n,$

the desired characteristic equation is:

$$q(s) = (s - s_1)(s - s_2)\cdots(s - s_n) = s^n + \alpha_{n-1}s^{n-1} + \cdots + \alpha_1 s + \alpha_0$$

$$[q(A)] = [A]^n + \alpha_{n-1}[A]^{n-1} + \cdots \alpha_1[A] + \alpha_0[I]$$

This formula requires the inverse of the controllability matrix and hence should have a nonzero determinant. So, the system should be controllable.

EXAMPLE

$$\left\{ \begin{array}{c} \dot{x}_1 \\ \dot{x}_2 \end{array} \right\} = \left[\begin{array}{cc} 0 & 1 \\ 0 & 0 \end{array} \right] \left\{ \begin{array}{c} x_1 \\ x_2 \end{array} \right\} + \left\{ \begin{array}{c} 0 \\ 1 \end{array} \right\} u$$

Design a controller using Ackermann's method for desired poles of $s_{1,2} = -1 \pm j$.

$$[K_1 \quad K_2] = [0 \quad 1][P_c]^{-1}[q(A)]$$

$$[P_c] = [\{P_1\} \quad \{P_2\}]$$

$$\{P_1\} = \{B\} = \begin{Bmatrix} 0 \\ 1 \end{Bmatrix}$$

$$\{P_2\} = [A]\{P_1\} = \begin{Bmatrix} 1 \\ 0 \end{Bmatrix}$$

$$[P_c] = \begin{bmatrix} 0 & 1 \\ 1 & 0 \end{bmatrix}; |P_c| = -1$$

$$[P_c]^{-1} = (-1)\begin{bmatrix} 0 & -1 \\ -1 & 0 \end{bmatrix} = \begin{bmatrix} 0 & 1 \\ 1 & 0 \end{bmatrix}$$

The desired characteristic equation from the given poles:

$$q(s) = (s - s_1)(s - s_2) = s^2 + 2s + 2$$

For a second-order system: $q(s) = s^2 + \alpha_1 s + \alpha_0$; comparing these two equations: $\alpha_1 = 2$, $\alpha_0 = 2$

$$[q(A)] = [A]^2 + \alpha_1[A] + \alpha_0[I] = \begin{bmatrix} 0 & 1 \\ 0 & 0 \end{bmatrix}\begin{bmatrix} 0 & 1 \\ 0 & 0 \end{bmatrix} + 2\begin{bmatrix} 0 & 1 \\ 0 & 0 \end{bmatrix} + 2\begin{bmatrix} 1 & 0 \\ 0 & 1 \end{bmatrix} = \begin{bmatrix} 2 & 2 \\ 0 & 2 \end{bmatrix}$$

$$[K_1 \quad K_2] = [0 \quad 1]\begin{bmatrix} 0 & 1 \\ 1 & 0 \end{bmatrix}\begin{bmatrix} 2 & 2 \\ 0 & 2 \end{bmatrix} = [2 \quad 2]$$

Observer Design

The design objective is to determine the observer gain matrix, [L], for the desired pole location. If a system is completely observable with a given set of outputs, it is possible to estimate all the states.

For a given state equation, $\{\dot{x}\} = [A]\{x\} + \{B\}u$, and an output equation, $y = [C]\{x\}$:

The full state observer for the system is given as, (Ref: D.G. Luenberger, "Observing the State of a Linear System", IEEE Transactions, 1964)

$$\{\dot{\hat{x}}\} = A\hat{x} + Bu + L(y - C\hat{x}); \quad \{L\} = \begin{Bmatrix} L_1 \\ L_2 \\ \vdots \\ L_n \end{Bmatrix}$$

where $\{\dot{\hat{x}}\}$ is the estimated value of states, $\{x\}$, and $\{L\}$ is the observer gain and a column matrix. The goal of observer is to provide an estimate, $\hat{x}$, so that $\hat{x} \to x$ as $t \to \infty$. Because we don't know $x(t_0)$ precisely, we must provide an initial estimate, $\hat{x}(t_0)$. Now, the error in estimation can be defined as $e(t) = x(t) - \hat{x}(t)$ and $e(t) \to 0$ as $t \to \infty$.

$$\dot{e} = \dot{x} - \dot{\hat{x}}$$

$$\dot{x} = Ax + Bu; \dot{\hat{x}} = A\hat{x} + Bu + L[Cx - C\hat{x}] = A\hat{x} + Bu + LCe$$

$$\dot{e} = A(x - \hat{x}) - LCe = Ae - LCe$$

$$\dot{e} = (A - LC)e$$

Let $e(t) = e^{st}$; $\dot{e} = se^{st}$:

$$se^{st} = (A - LC)e^{st}$$

$$e^{st}[sI - (A - LC)] = 0$$

So, the determinant is $\left| sI - (A - LC) \right| = 0$.

EXAMPLE
Given that:

$$\begin{Bmatrix} \dot{x}_1 \\ \dot{x}_2 \end{Bmatrix} = \begin{bmatrix} 2 & 3 \\ -1 & 4 \end{bmatrix} \begin{Bmatrix} x_1 \\ x_2 \end{Bmatrix} + \begin{Bmatrix} 0 \\ 1 \end{Bmatrix} u; \text{ and } y = \begin{bmatrix} 1 & 0 \end{bmatrix} \begin{Bmatrix} x_1 \\ x_2 \end{Bmatrix}$$

Design an observer so that the poles are $s_{1,2} = -8 \pm 6j$.

$$LC = \begin{Bmatrix} L_1 \\ L_2 \end{Bmatrix} \begin{bmatrix} 1 & 0 \end{bmatrix} = \begin{bmatrix} L_1 & 0 \\ L_2 & 0 \end{bmatrix}$$

$$[A - LC] = \begin{bmatrix} (2 - L_1) & 3 \\ (-1 - L_2) & 4 \end{bmatrix}$$

$$s[I] - [A - LC] = \begin{bmatrix} s - (2 - L_1) & -3 \\ (1 + L_2) & s - 4 \end{bmatrix}$$

The determinant, $\left| sI - (A - LC) \right|$, is:

$$s^2 + s(L_1 - 6) + (3L_2 - 4L_1 + 11) = 0 \rightarrow (1)$$

For the given poles, the desired characteristic equation is:

$$(s - s_1)(s - s_2) = 0$$

$$s^2 - (s_1 + s_2)s + s_1 s_2 = 0$$

$$s^2 + 16s + 100 = 0 \rightarrow (2)$$

Comparing equations (1) and (2):

$$L_1 - 6 = 16; \ L_1 = 22$$

$$3L_2 - 4L_1 + 11 = 100; \ L_2 = 177/3 = 59$$

The observer gain matrix:

$$\begin{Bmatrix} L_1 \\ L_2 \end{Bmatrix} = \begin{Bmatrix} 22 \\ 59 \end{Bmatrix}$$

Comparing equation (2) with the standard form:

$$s^2 + 2\zeta w_n s + w_n^2 = 0$$

$$w_n = 10, \ \zeta = 0.8$$

From these values, the correct PO and T_s can be determined.

Ackermann's Formula Method

For the nth-order system:

$$\{L\} = \begin{Bmatrix} L_1 \\ L_2 \\ \vdots \\ L_n \end{Bmatrix}$$

Then, Ackermann's formula for the observer gain matrix is:

$$\{L\} = [p(A)][P_0]^{-1} \begin{Bmatrix} 0 \\ 0 \\ \vdots \\ 1 \end{Bmatrix}$$

where $[P_0]$ is the observability matrix.

If the desired poles are $s_1, s_2, \cdots s_n$:

The desired characteristic equation is:

$$p(s) = s^n + \beta_{n-1}s^{n-1} + \cdots + \beta_1 s + \beta_0$$

$$[p(A)] = [A]^n + \beta_{n-1}[A]^{n-1} + \cdots + \beta_1[A] + \beta_0[I]$$

This formula requires the inverse of the observability matrix and hence should have a nonzero determinant. So, the system should be observable.

EXAMPLE

Given that:

$$[A] = \begin{bmatrix} 2 & 3 \\ -1 & 4 \end{bmatrix}; \{B\} = \begin{Bmatrix} 0 \\ 1 \end{Bmatrix}; [C] = [\ 1 \quad 0\ ]$$

Design an observer so that the poles are $s_{1,2} = -8 \pm 6j$.

$$[P_0] = \begin{bmatrix} [p_1] \\ [p_2] \end{bmatrix}$$

$$[P_1] = [C] = [\ 1 \quad 0\ ]$$

$$[P_2] = [P_1][A] = [\ 2 \quad 3\ ]$$

$$[P_0] = \begin{bmatrix} 1 & 0 \\ 2 & 3 \end{bmatrix}; |P_0| = 3; \text{system is observable.}$$

$$[P_0]^{-1} = \left(\frac{1}{3}\right)\begin{bmatrix} 3 & 0 \\ -2 & 1 \end{bmatrix} = \begin{bmatrix} 1 & 0 \\ -2/3 & 1/3 \end{bmatrix}$$

For the given poles $(s_1 = -8 + 6j, \quad s_2 = -8 - 6j)$:

The desired characteristic equation: $p(s) = s^2 - (s_1 + s_2)s + s_1 s_2 = s^2 + 16s + 100$

$$[p(A)] = [A]^2 + 16[A] + 100[I]$$

$$[P(A)] = \begin{bmatrix} 2 & 3 \\ -1 & 4 \end{bmatrix}\begin{bmatrix} 2 & 3 \\ -1 & 4 \end{bmatrix} + 16\begin{bmatrix} 2 & 3 \\ -1 & 4 \end{bmatrix} + 100\begin{bmatrix} 1 & 0 \\ 0 & 1 \end{bmatrix} = \begin{bmatrix} 133 & 66 \\ -22 & 177 \end{bmatrix}$$

$$\begin{Bmatrix} L_1 \\ L_2 \end{Bmatrix} = [p(A)[P_0]]^{-1}\begin{Bmatrix} 0 \\ 1 \end{Bmatrix} = \begin{bmatrix} 133 & 66 \\ -22 & 177 \end{bmatrix}\begin{bmatrix} 1 & 0 \\ -2/3 & 1/3 \end{bmatrix}\begin{Bmatrix} 0 \\ 1 \end{Bmatrix} = \begin{Bmatrix} 22 \\ 59 \end{Bmatrix}$$

The observer gain matrix: $\begin{Bmatrix} L_1 \\ L_2 \end{Bmatrix} = \begin{Bmatrix} 22 \\ 59 \end{Bmatrix}$

10.10. Summary

In this chapter, for a given system with an nth-order differential equation, it was shown how to identify the state variables and reduce the system into a state variable system represented by the state and output equations. For the given initial conditions, the state equation can be easily solved by the LT to find the system's response in the time domain. Similarly, for a given TF, the state variables can be identified, and the resulting state variable system can be represented by the state and output equations. The state variable system can also be modeled using the block diagram method or the signal flow graph method. The characteristic equation can be easily derived from the given state equation and used to study the system's stability by the R-H criterion. The state variable system design consists of the design of a controller and an observer. However, a method was included to check the system first for controllability and observability. Then, it was demonstrated how to design a controller and an observer by a direct method as well as using Ackermann's formula method.

10.11. Assessment

1. A fourth-order system with state variables reduces to:

   a. Two second-order systems

   b. One third-order and one first-order system

   c. Four first-order systems

   d. None of the above

2. The size of [A] matrix in the state equation of a third-order system is:

   a. 3×3

   b. 3×1

   c. 1×1

   d. 1×3

3. The size of the [C] matrix in the output equation of a third-order system is:

   a. 3×3

   b. 3×1

   c. 1×1

   d. 1×3

4. If the TF of a system is given, the state equation can be obtained from the:

   a. Denominator polynomial

   b. Input function in the time domain

   c. Numerator polynomial

   d. Output function in the time domain

5. The output equation can be obtained from the:

    a. Denominator polynomial

    b. Input function in the time domain

    c. Numerator polynomial

    d. Output function in the time domain

6. A system is controllable if:

    a. $[P_O] = 0$

    b. $[P_C] = 0$

    c. $[P_O] \neq 0$

    d. $[P_C] \neq 0$

7. A system is observable if:

    a. $[P_O] = 0$

    b. $[P_C] = 0$

    c. $[P_O] \neq 0$

    d. $[P_C] \neq 0$

8. Ackermann's formula used in design requires:

    a. $[P_O] \neq 0$ and $[P_C] \neq 0$

    b. $[P_O] = 0$ and $[P_C] \neq 0$

    c. $[P_O] = 0$ and $[P_C] = 0$

    d. $[P_O] \neq 0$ and $[P_C] = 0$

9. Controller gain in Ackermann's formula is a:

    a. Column matrix

    b. Square matrix

    c. Row matrix

    d. Single value

10. Observer gain in Ackermann's formula is a:

    a. Column matrix

    b. Square matrix

    c. Row matrix

    d. Single value

10.12. Practice Problems

1. Answer all the assessment questions in Section 10.11.

2. For an autonomous car, the transfer function is given as,

$$T(s) = \frac{2s^2 + 6s + 5}{(s+1)(s^2 + 3s + 2)}$$

    Determine the State equation and the Output equation.

3. For the above problem, draw the (i) Block diagram, and (ii) Signal Flow Graph diagram

4. For the state equation and output equation obtained in problem (2), determine the system response in time domain. Assume unit step input and the initial conditions for each state variable is equal to one.

5. State variable model of a system is given as, $\{\dot{x}\} = [A]\{x\}$, where

$$[A] = \begin{bmatrix} 0 & 1 & 0 \\ 0 & 0 & 1 \\ -6 & -1 & -3K \end{bmatrix}$$

    Find the value of K, if the system is marginally stable.

6. Given, $\{\dot{x}\} = [A]\{x\} + [B]\{u\}$, where $\{u\} = [K]\{x\}$; The matrices, A, B, and K are

$$[A] = \begin{bmatrix} 0 & 1 \\ -1 & 0 \end{bmatrix}; \quad [B] = \begin{bmatrix} 1 & 0 \\ 0 & 1 \end{bmatrix}; \quad [K] = \begin{bmatrix} -k & 0 \\ 0 & -2k \end{bmatrix}$$

    Determine the value of "k" such that the system is critically damped ($\zeta = 1$)

7. Given the state equation as, $\{\dot{x}\} = [A]\{x\} + [B]\{u\}$, where the matrices A and B are

$$[A] = \begin{bmatrix} 0 & 1 \\ 9 - k1 & -k2 \end{bmatrix}; \quad [B] = \begin{bmatrix} 0 \\ 1 \end{bmatrix}$$

Determine the value of kl and $k2$, if the settling time is 2 sec, and the system is critically damped $(\zeta = 1)$.

8. Given the state equation and output equation as given below, determine if the system is controllable and observable.

$\{\dot{x}\} = [A]\{x\} + [B]\{u\}; y = [C]\{x\}$; The matrices, A, B, C are:

$$[A] = \begin{bmatrix} 0 & 1 \\ 0 & -5 \end{bmatrix}; \quad [B] = \begin{bmatrix} 0 \\ 2 \end{bmatrix}; \quad [C] = [\,0 \quad 2\,]$$

9. Given the state equation and output equation as given below, determine if the system is controllable and observable.

$\{\dot{x}\} = [A]\{x\} + [B]\{u\}; y = [C]\{x\}$; The matrices, A, B, C are:

$$[A] = \begin{bmatrix} 0 & 1 \\ -2 & -2 \end{bmatrix}; \quad [B] = \begin{bmatrix} 1 \\ -3 \end{bmatrix}; \quad [C] = [\,1 \quad 0\,]$$

10. Given the state equation and output equation as given below,

$\{\dot{x}\} = [A]\{x\} + [B]\{u\}; y = [C]\{x\}$; The matrices, A, B, C are:

$$[A] = \begin{bmatrix} -4 & 0 \\ 1 & -1 \end{bmatrix}; \quad [B] = \begin{bmatrix} 1 \\ 0 \end{bmatrix}; \quad [C] = [\,0 \quad 1\,]$$

Design a controller by Ackermann's method such that the desired poles are $S_1 = -1 + 3j$; $S_2 = -1 - 3j$

Appendix A

MATLAB

MATLAB is a tool for engineering computations, and it is based on matrices. For problems related to control systems, MATLAB has a collection of special files (M-files) in the "Control System Toolbox".

Writing a Matrix in MATLAB

$$[A] = \begin{bmatrix} 1 & 2 & 3 \\ 6 & 7 & 8 \\ 11 & 12 & 13 \end{bmatrix}$$

In MATLAB, write row by row >> $A = [1\ 2\ 3;\ 6\ 7\ 8;\ 11\ 12\ 13]$

">>" denotes the command prompt on the screen. Each row is separated by a semi-colon.

Writing a Vector

If a vector starts from 10 and increases in steps of 5 up to 50, it can be written as:

$$\gg x = \text{linspace}(n1, n2, n)$$

$$= \text{linspace}(10,\ 50,\ 5)$$

$$x = 10\ 15\ 20\ 25\ 30\ 35\ 40\ 45\ 50$$

Addition and Subtraction

$$\gg A = [2\ 3;\ 4\ 5;\ 6\ 7] \Rightarrow \begin{bmatrix} 2 & 3 \\ 4 & 5 \\ 6 & 7 \end{bmatrix}$$

$$\gg B = [1\ 0;\ 3\ 4;\ 0\ 5] \Rightarrow \begin{bmatrix} 1 & 0 \\ 3 & 4 \\ 0 & 5 \end{bmatrix}$$

$$\gg C = A + B \Rightarrow C = \begin{matrix} 3 & 3 \\ 7 & 9 \\ 6 & 12 \end{matrix}$$

$$\gg C = A - B \Rightarrow \begin{matrix} 1 & 3 \\ 1 & 1 \\ 6 & 2 \end{matrix}$$

Multiplication

$$\gg X = [1;\ 3;\ 5] \Rightarrow X = \begin{matrix} 1 \\ 3 \\ 5 \end{matrix}$$

$$\gg Y = [2;\ 4;\ 6] \Rightarrow Y = \begin{matrix} 2 \\ 4 \\ 6 \end{matrix}$$

$$\gg Z = X' * Y \Rightarrow \begin{bmatrix} 1 & 3 & 5 \end{bmatrix} \begin{bmatrix} 2 \\ 4 \\ 6 \end{bmatrix} = 44$$

$$\gg Z = X * Y' \Rightarrow \begin{bmatrix} 1 \\ 3 \\ 5 \end{bmatrix} \begin{bmatrix} 2 & 4 & 6 \end{bmatrix} = \begin{matrix} 2 & 4 & 6 \\ 6 & 12 & 18 \\ 10 & 20 & 30 \end{matrix}$$

A Vector With a Complex Number

$$\gg Y = [\ 6+5*i \quad 3+4*i \quad 1-i\ ]; \Rightarrow [\ 6+5j \quad 3+4j \quad 1-j\ ]$$

$$\gg Y.^2 \Rightarrow [\ 11+60i \quad -7+24i \quad 0-2i\ ]$$

Roots of Characteristic Equation

$$s^3 + 6s^2 + 11s + 6 = 0$$

$$\gg p = [1 \quad 6 \quad 11 \quad 6];$$

$$\gg r = \text{roots}(p)$$

$$r = \begin{matrix} -3 \\ -2 \\ -1 \end{matrix} \quad \Rightarrow s_1 = -1, \quad s_2 = -2, \quad s_3 = -3$$

$$s^3 + 6s^2 + 11s + 6 = (s-s_1)(s-s_2)(s-s_3) = (s+1)(s+2)(s+3)$$

Given the Roots, Find the Characteristic Equation

$$\gg r = [\ -3 \quad -2 \quad -1\ ];$$

$$\gg q = \text{poly}(r)$$

$$q = \quad 1 \quad 6 \quad 11 \quad 6$$

$$q = s^3 + 6s^2 + 11s + 6$$

Eigenvalues

$$[A] = \begin{bmatrix} 0 & 1 & 0 \\ -1 & 0 & 2 \\ 3 & 0 & 5 \end{bmatrix}$$

$$\gg A = [0 \quad 1 \quad 0; \ -1 \quad 0 \quad 2; \ 3 \quad 0 \quad 5]$$

$$\gg \text{eig}[A] \Rightarrow \text{Eigenvalues:} \quad \begin{matrix} 5.213 \\ -0.1065 + 1.4487i \\ -0.1065 - 1.4487i \end{matrix}$$

Eigenvalues and Eigenvectors

$$[A] = \begin{bmatrix} 0 & 1 & 0 \\ 0 & 0 & 1 \\ -6 & -11 & -6 \end{bmatrix}$$

$$\gg A = [0 \ \ 1 \ \ 0; \ \ 0 \ \ 0 \ \ 1; \ \ -6 \ \ -11 \ \ -6];$$

$$\gg [X, D] = \text{eig}(A)$$

$$\text{Eigenvalues: } D = \begin{matrix} -1 & 0 & 0 \\ 0 & -2 & 0 \\ 0 & 0 & -3 \end{matrix}$$

$$\text{Eigen vector: } X = \begin{matrix} -0.5774 & 0.2182 & -0.1048 \\ 0.5774 & -0.4364 & 0.3145 \\ -0.5774 & 0.8729 & -0.9435 \end{matrix}$$

Inverse of a Matrix

$$[A] = \begin{bmatrix} 1 & 1 & 2 \\ 3 & 4 & 0 \\ 1 & 2 & 5 \end{bmatrix}$$

$$\gg A = [1 \ \ 1 \ \ 2; \ \ 3 \ \ 4 \ \ 0; \ \ 1 \ \ 2 \ \ 5];$$

$$\gg \text{inv}(A)$$

$$[A]^{-1} = \begin{bmatrix} 2.2222 & -0.1111 & -0.8889 \\ -1.6667 & 0.3333 & 0.6667 \\ 0.2222 & -0.1111 & 0.1111 \end{bmatrix}$$

NOTE: All function names MUST be in LOWER CASE.

Example: inv(A), eig(A), poly(A) etc.

Partial Fraction

$$\frac{B(s)}{A(s)} = \frac{r_1}{s - p_1} + \frac{r_2}{s - p_2} + \cdots + \frac{r_n}{s - p_n} + k$$

Example:

$$\frac{B(s)}{A(s)} = \frac{2s^3 + 5s^2 + 3s + 6}{s^3 + 6s^2 + 11s + 6}$$

```
>> num = [2   5   3   6];
>> den = [1   6   11   6];
>> [r, p, k] = residue(num, den)
```

$$r =$$
```
    -6.0000
    -4.0000
     3.0000
```

$$p =$$
```
    -3.0000
    -2.0000
    -1.0000
```

$$k =$$
```
     2
```

$$\frac{B(s)}{A(s)} = \frac{-6}{s + 3} + \frac{-4}{s + 2} + \frac{3}{s + 1} + 2$$

Given the above relationship, the original function can be obtained from:

```
>> r = [-6   -4   3];
>> p = [-3   -2   -1];
>> k = 2;
>> [num, den] = residue(r, p, k);
>> printsys (num, den, 's')
num/den =
```

$$\frac{2s^3 + 5s^2 + 3s + 6}{s^3 + 6s^2 + 11s + 6}$$

Poles and Zeros of a Transfer Function

$$T(s) = \frac{4s^2 + 16s + 12}{s^4 + 12s^3 + 44s^2 + 48s}$$

MATLAB command:

$$[z, p, k] = \text{tf2zp (num, den)}$$

```
>> num = [4   16   12];
>> den = [1   12   44   48   0];
>> [z, p, k] = tf2zp(num, den)
```

z =
       −3
       −1
p =
           0
      −6.0000
      −4.0000
      −2.0000

k =
       4

'z', 'p', 'k' refer to zeros, poles, gain, respectively $z_2 \approx -3$, $z_1 \approx -1$.

$$T(s) = (4)\frac{(s+1)(s+3)}{s(s+2)(s+4)(s+6)}$$

The command to find $T(s)$ from a given z, p, k:

```
zp2tf(z, p, k)
>> z = [−1;   −3];
>> p = [0;   −2;   −4;   −6];
>> k = 4;
>> [num, den] = zp2tf(z, p, k);
>> printsys(num, den, 's')
```

num/den =

$$\frac{4s^2 + 16s + 12}{s^4 + 12s^3 + 44s^2 + 48s}$$

System Response

Let the transfer function for a second-order system be:

$$T(s) = \frac{2(s+5)}{s^2 + 2s + 10} = \frac{Y(s)}{R(s)}$$

For unit step input: $R(s) = U(s)$

$$\frac{Y(s)}{U(s)} = \frac{2s + 10}{s^2 + 2s + 10}$$

Let the time interval for response $0 \leq t \leq 8$ be in steps of every 0.01 second.

```
>> t = 0:0.01:8;
>> num = [2   10];
>> den = [1   2   10];
>> sys = tf(num, den);
>> y = step(sys, t);
>> plot(t, y); grid
```

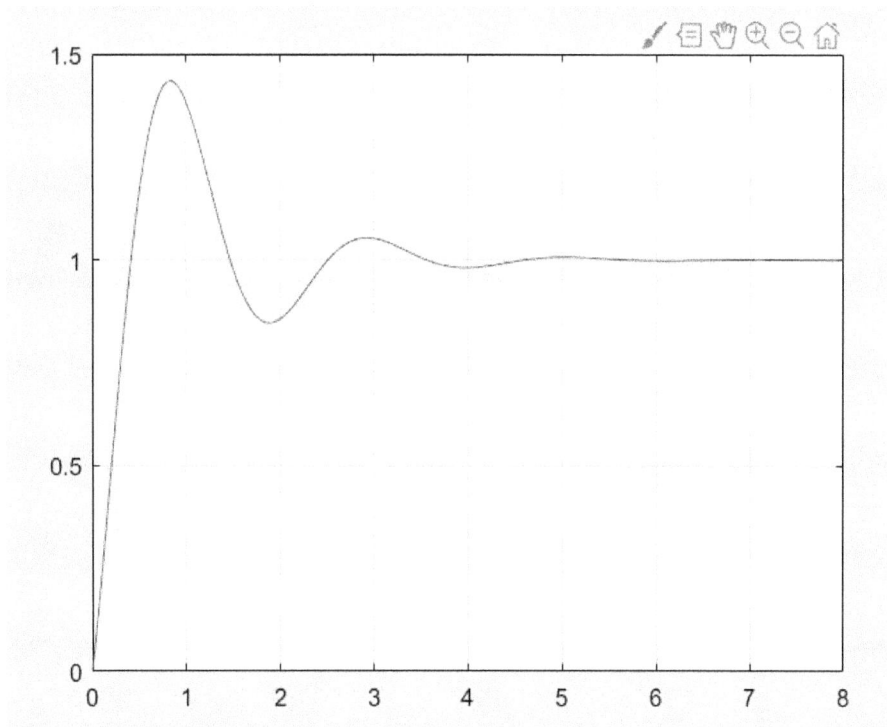

Image A.11

Performance Parameters

(Rise time, peak time, percent overshoot, settling time)

$$T(s) = \frac{25}{s^2 + 6s + 25}$$

Step response for $0 \leq t \leq 5$:

Time interval $\Delta t = 0.005$ sec:

```
1-   t = 0: 0.005: 5;
2-   num = [25];
3-   den = [1   6   25];
4-   [y, x, t] = step(num, den, t);
5-   r = 1;
6-   while y(r) < 1.0001
7-       r = r + 1;
8-   end
9-   rise_time = (r − 1)*0.005;
10-  [ymax, tp] = max(y);
11-  peak_time = (tp − 1)*0.005;
12-  max_overshoot = ymax − 1;
13-  s = 1001;
14-  while y(s) > 0.98 && y(s) < 1.02
15-      s = s − 1;
16-  end
17-  settling_time = (s − 1)*0.005;
```

Create a new script in MATLAB then type the codes. After running it, the results can be found in workspace. Correct answer is showing below

settling_time =

1.1850

rise_time =

0.5550

peak_time =

0.7850

max_overshoot =

0.0948

Impulse and Ramp Responses as a Function of Step Response

Let $\dfrac{Y(s)}{R(s)} = \dfrac{1}{s^2 + 0.2s + 1} = T(s)$

$$\text{Impulse: } R(s) = 1$$

$$\text{Step: } R(s) = 1/s$$

Impulse response $= \dfrac{1}{s}[sT(s)] = $ step response of $[sT(s)]$

$$T_1(s) = sT(s) = \dfrac{s}{s^2 + 0.2s + 1}$$

```
>> num = [1   0];
>> den = [1   0.2   1];
>> y = step(num, den);
>> plot(y)
>>
```

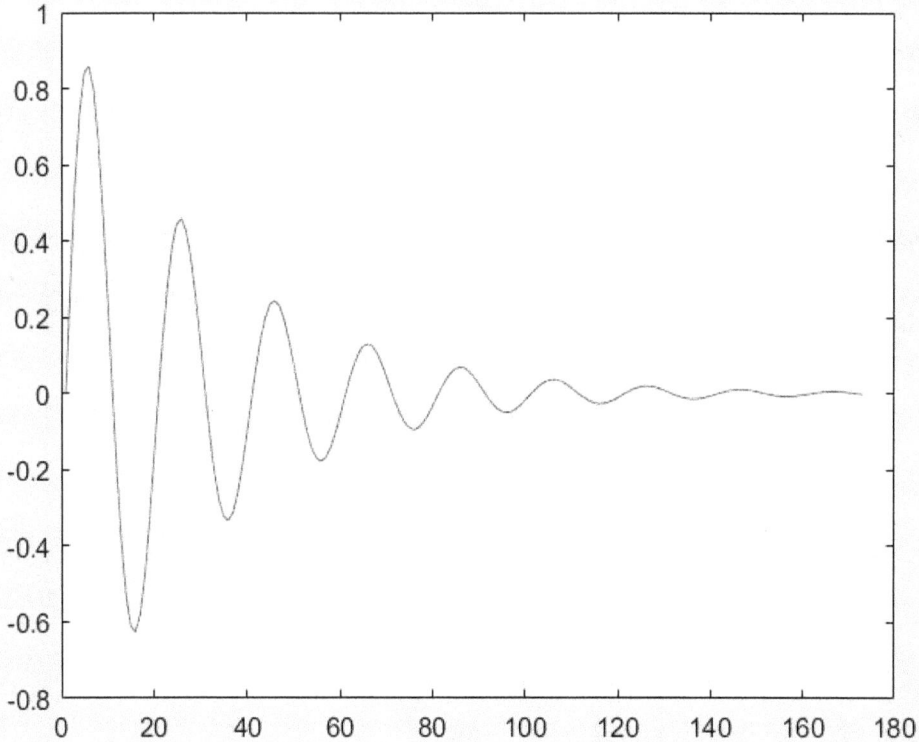

Ramp: $R(s) = \dfrac{1}{s^2}$, Ramp response $= \dfrac{1}{s}\left[\dfrac{1}{s}T(s)\right] =$ step response of $\left[\dfrac{1}{s}T(s)\right]$

Let $T(s) = \dfrac{1}{s^2+s+1}; 0 \le t \le 7; \Delta t = 0.1$

$$T_1(s) = \frac{1}{s}[T(s)] = \frac{1}{s^3+s^2+s}$$

```
>> num = [1];
>> den = [1   1   1   0];
>> t = 0: 0.1: 7;
>> y = step(num, den, t);
>> plot(t, y); grid
>>
```

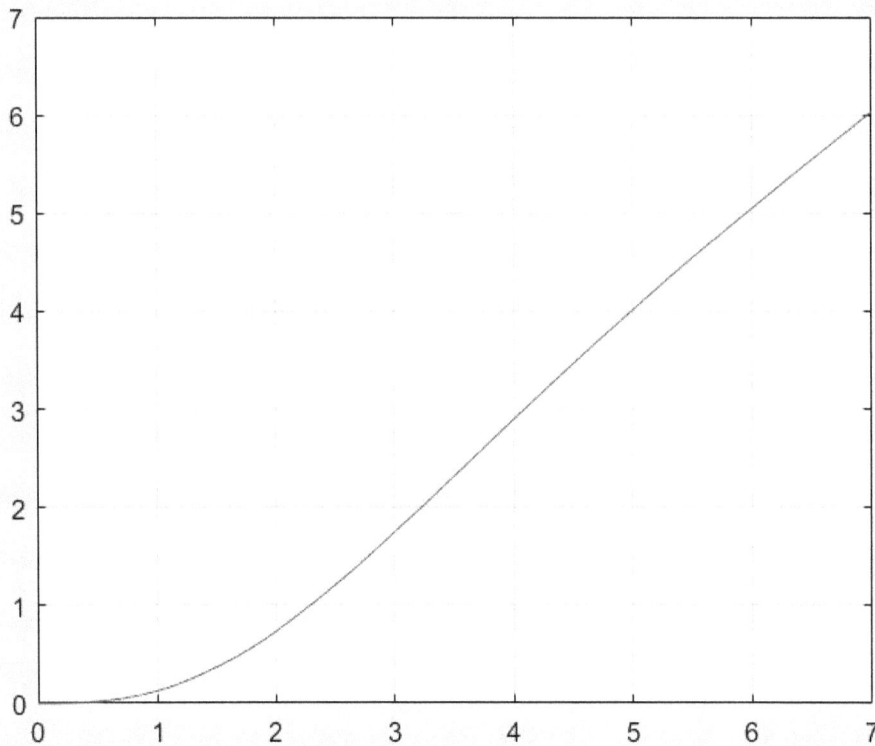

Image A.13

Dynamic System With Initial Conditions

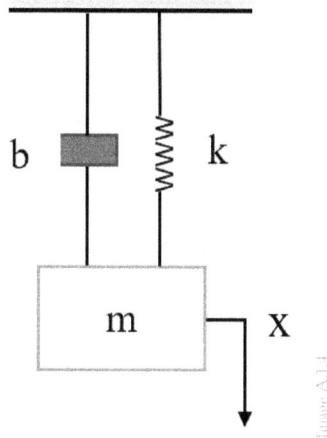

Initial conditions:

$$x(0) = x_0, \quad \dot{x}_0(0) = v_0$$

Given:

$$m = 1kg, \, b = 3kg/s, \, k = 2N/m, \, x_0 = 0.1m, \, v_0 = 0.05m/s$$

Equation of motion: $m\ddot{x} + b\dot{x} + kx = 0$

The Laplace transform of the above equation gives:

$$X(s) = \frac{(ms + b)x_0 + mv_0}{ms^2 + bs + k} = \frac{(s + 3)(0.1) + (1)(0.05)}{s^2 + 3s + 2} = \frac{0.1s + 0.35}{s^2 + 3s + 2}$$

Response $= \frac{1}{s}[sX(s)] =$ step response of $[sX(s)]$

Let $X_1(s) = sX(s) = \frac{0.1s^2 + 0.35s}{s^2 + 3s + 2}$

Step response of $X_1(s)$

```
>> num = [0.1   0.35   0];
>> den = [1   3   2];
>> step(num, den); grid
```

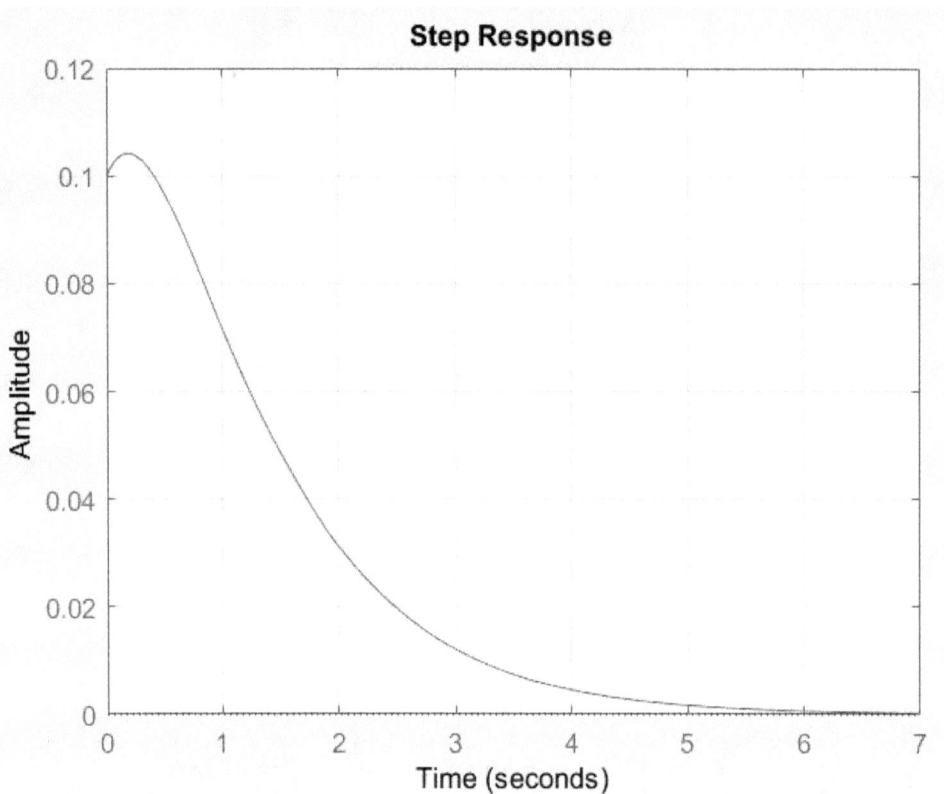

Root Locus

Commands that are useful:

*"**conv**"*: convolution (To multiply polynomials)

*"**deconv**"*: deconvolution (To divide polynomials)

*"**rlocus**"*: To plot a root locus diagram

Example

Transfer Function

$$\frac{K(s+3)}{s(s+1)(s^2+4s+16)}, \quad K=1$$

The range for x and y axes: $-6 \le x \le 6, -6 \le y \le 6$

Zero: $s = -3$ Poles: Multiply $s(s+1)(s^2+4s+16)$ and then find the roots.

Let $a = s^2 + s, b = s^2 + 4s + 16$

```
>> a = [1  1  0];
>> b = [1  4  16];
>> c = conv (a, b)


c =

    1   5   20   16   0

>> r = roots (b)


r =
    -2.0000 + 3.4641i
    -2.0000 - 3.4641i
```

$$c = s^4 + 5s^3 + 20s^2 + 16s$$

$$\text{Roots of } s^2 + 4s + 16$$

$$\text{Poles : } 0, -1, -2 \pm 3.4641i$$

$$\text{Zero : } -3$$

Root Locus Diagram

```
num = [1  3];
den = [1  5  20  16  0];
r = rlocus(num, den);
plot (r, '_');
v = [-6  6  -6  6]; axis(v)
grid
xlabel('Real Axis');
ylabel('Imag Axis');
gtext('o') % place 'o' mark on open loop zero
gtext('x') % place 'x' mark on each pole
gtext('x')
gtext('x')
gtext('x')
```

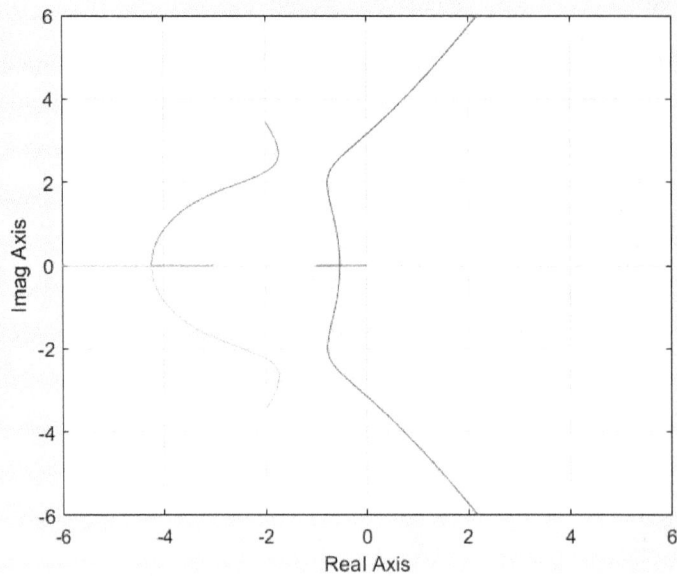

Bode Plot

$$G(s) = \frac{25}{s^2 + 4s + 25}$$

```
>> num = [25];
>> den = [1   4   25];
>> bode(num, den); grid
```

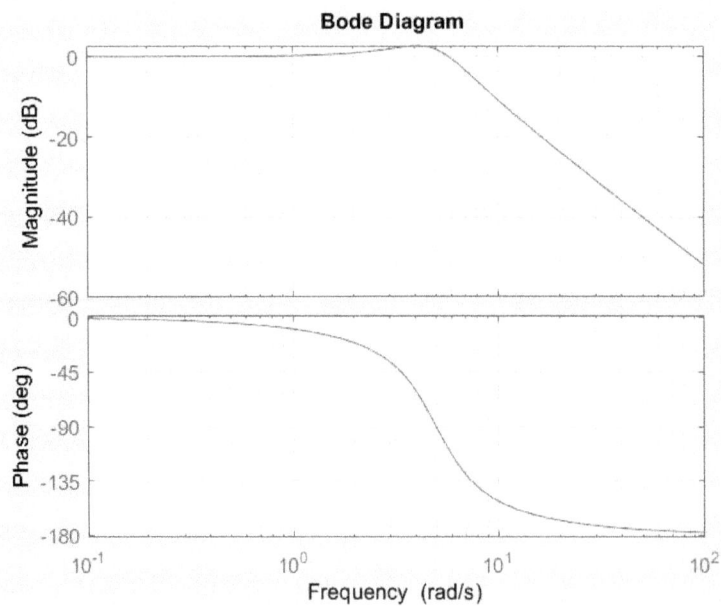

Nyquist Plot

$$G(s) = \frac{1}{s^2 + 0.8s + 1}$$

```
>> num = [1];
>> den = [1   0.8   1];
>> nyquist(num, den); title ('Nyquist')
>>
```

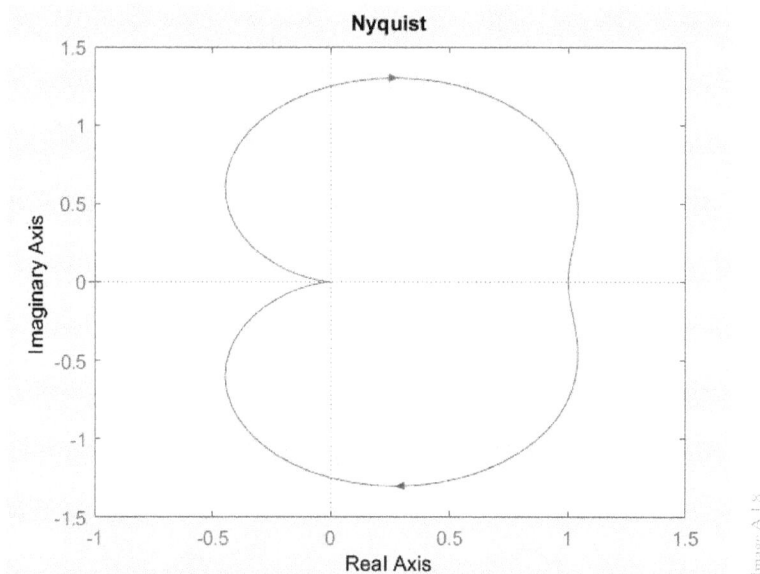

Phase Margin and Gain Margin

These are measurements of closeness of the polar plot to $(-1 + j0)$ point.

Gm: Gain Margin

Pm: Phase Margin

wcp: Phase cross over frequency (ω_o)

wcg: Gain cross over frequency (ω_c)

Command: [Gm Pm wcp wcg] = margin(sys)

Example

$$G(s) = \frac{20(s + 1)}{s(s + 5)(s^2 + 2s + 10)}$$

```
num = [20   20];
den = conv([1   5   0], [1   2   10]);
sys = tf (num, den);
w = logspace(−1, 2, 100);
bode(sys, w)
[Gm, Pm, wcp, wcg] = margin(sys);
GmdB = 20*log10(GM);
[GmdB Pm wcp wcg];
```

Bode Diagram

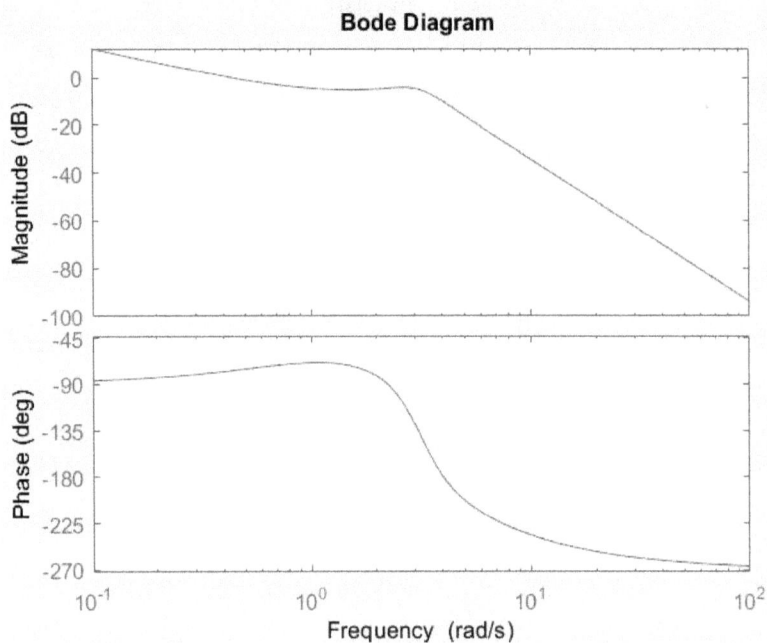

ans =

    9.9301   103.6573   4.0132   0.4426

State-Space Control

Controllability and Observability

$$CONT = ctrb(A, B)$$

$$OBSER = obsv(A, C)$$

$$\{\dot{x}\} = [A]\{x\} + \{B\}u$$

$$y = [C]\{x\}$$

EXAMPLE

Given: $[A] = \begin{bmatrix} 0 & 1 & 0 \\ 0 & 0 & 1 \\ -6 & -11 & -6 \end{bmatrix}$ $\{B\} = \begin{Bmatrix} 0 \\ 0 \\ 1 \end{Bmatrix}$ $[C] = \begin{bmatrix} 5 & 6 & 1 \end{bmatrix}$

\>> $A = \begin{bmatrix} 0 & 1 & 0; & 0 & 0 & 1; & -6 & -11 & -6 \end{bmatrix}$;
\>> $B = \begin{bmatrix} 0; & 0; & 1 \end{bmatrix}$;
\>> $C = \begin{bmatrix} 5 & 6 & 1 \end{bmatrix}$;
\>> $CONT = ctrb(A, B)$

CONT =

$$\begin{matrix} 0 & 0 & 1 \\ 0 & 1 & -6 \\ 1 & -6 & 25 \end{matrix}$$

\>> det(CONT)   System is controllable since the det (CONT) = -1

ans =

-1

\>> OBSER = obsv(A, C)

OBSER =

$$\begin{matrix} 5 & 6 & 1 \\ -6 & -6 & 0 \\ 0 & -6 & -6 \end{matrix}$$

\>> det(OBSER)   System is NOT observable since the det (OBSER) = 0

ans =

$-1.5987e-14$

MATLAB COMMANDS

Command	Description
abs	Absolute value, magnitude of complex number
angle	Phase angle of a complex number
atan	Arctangent $[\tan^{-1}(\ )]$
axis	Manual axis scaling
bode	Bode plot
clear	Clear workspace
clf	Clear current figure
conv	Convolution, Multiplication of polynomials
cos	Cosine
cosh	Hyperbolic Cosine
ctrb	Controllability Matrix computation
deconv	Deconvolution, Division of polynomials
det	Determinant
eig	Eigenvalues and eigenvectors
freqs	Laplace Transform frequency response
grid	Toggles the major lines of current axes
help	Lists all help topics
imag	Imaginary part
impulse	Impulse response
inv	Inverse of a matrix
log	Natural logarithm
Log10	Log base 10
logspace	Logarithmically spaced vector
lsim	Simulate time response
nyquist	Nyquist Plot
obsv	Compute Observability matrix
plot	Linear x-y Plot
poly	Compute polynomial from roots

Command	Description
residue	Partial fraction
rlocus	Root Locus Plot
roots	Compute roots from polynomial
sin	Sine function
sinh	Hyperbolic sine
sqrt	Square root
step	Unit-step response
sum	Addition
tan	Tangent
tanh	Hyperbolic Tangent
text	To position Text arbitrarily
tf	To find transfer function
title	Plot title
xlabel	X-axis label
ylabel	Y-axis label

CREDITS

Appendix A2

SIMULINK

Simulink is a simulation tool to study control systems. The graphical output helps visually see a system's behavior. The effect of changing the control gain or transfer function parameters can be easily observed through rapidly changing the output in this simulation model. It is very useful in tuning the control system to satisfy the desired performance parameters.

Model Building

The block diagram of a control system can be simulated with blocks available in the Simulink library. Some of these useful blocks are given below.

Sources

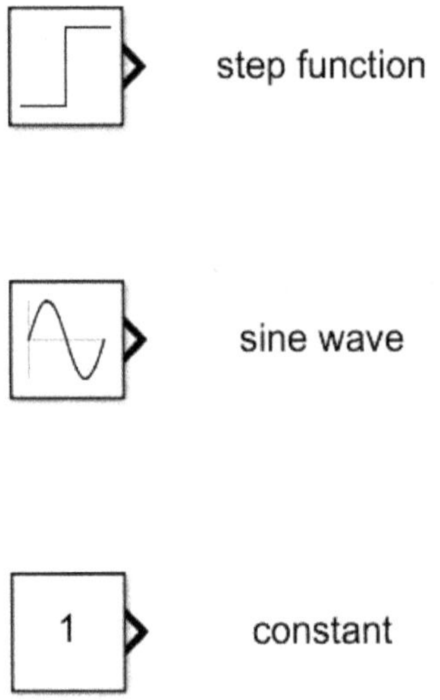

step function

sine wave

constant

Image A.2.1

Sink

scope

Image A.2.2

A block has a connecting point on the right or left side or on both sides, as shown below.

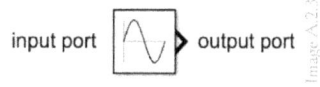

input port [block] output port

The input port is the connecting point on the left, and it brings input to the block. The output port is the connecting point on the right, and it takes the output to another block or scope.

Commonly used blocks:

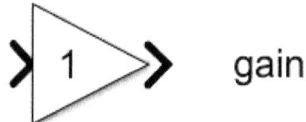

$$+ \atop +$$ sum

1   gain

Continuous blocks:

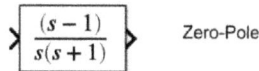

$\dfrac{1}{s+1}$ Transfer Function

$\dfrac{(s-1)}{s(s+1)}$ Zero-Pole

$\dfrac{1}{s}$ Integrator

$\dfrac{\Delta u}{\Delta t}$ Derivative

Working With Blocks

Double-click on summing block and change the feedback input sign from "+" to "-"

Double-click on gain block and enter the gain value (in this case enter 10)

Right click on the block and choose ROTATE&FLIP to rotate/flip it

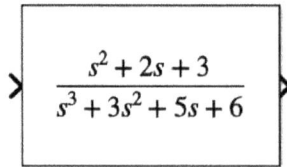

$$\frac{s^2 + 2s + 3}{s^3 + 3s^2 + 5s + 6}$$

To enter this transfer function, double-click on Transfer function block and enter numerator and denominator polynomial cofficients on the windows

---

Block Parameters: Transfer Fcn1   ✕

Transfer Fcn

The numerator coefficient can be a vector or matrix expression. The denominator coefficient must be a vector. The output width equals the number of rows in the numerator coefficient. You should specify the coefficients in descending order of powers of s.

Parameters

Numerator coefficients:

[1 2 3]

Denominator coefficients:

[1 3 5 6]

Absolute tolerance:

auto

State Name: (e.g., 'position')

"

| OK | Cancel | Help | Apply |

If poles and zeros are known, the zero-pole block can be used to simulate the transfer function.

Let and

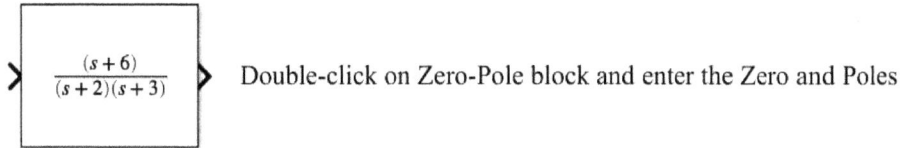

$$\frac{(s+6)}{(s+2)(s+3)}$$

Double-click on Zero-Pole block and enter the Zero and Poles

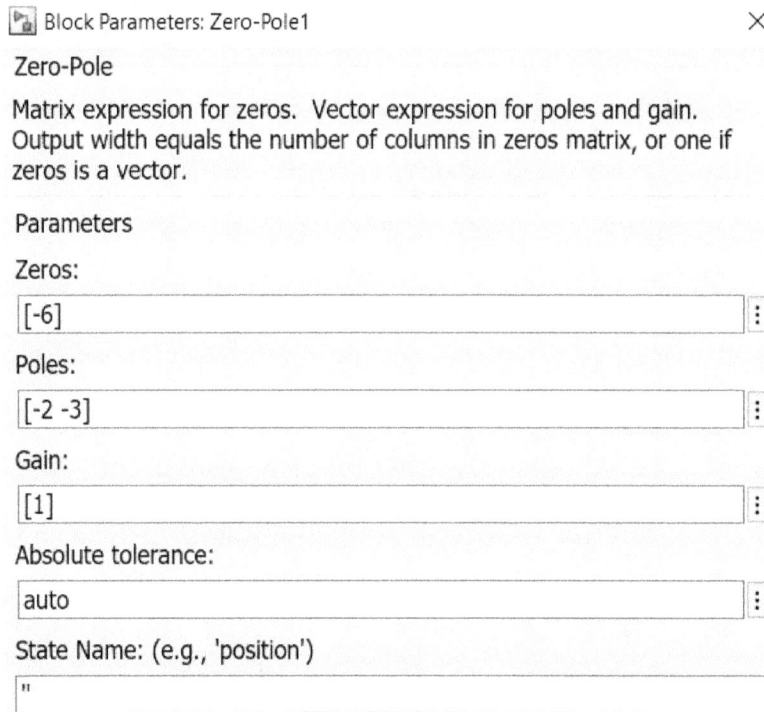

Forcing Functions on Simulink

Step:

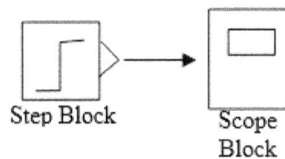

Step Block   Scope Block

(Parameters)

(0.01, 0, 1, 0)

Ramp:

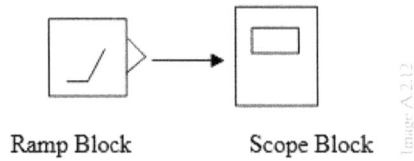

Ramp Block   Scope Block

(Slope $= 1$); change the start time to 0.01 on the ramp block to start the function from zero.

(Or)   step block and integrator block

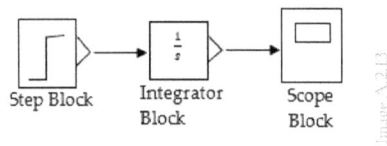

Step Block   Integrator   Scope
               Block   Block

Step block: (0.01, 0, 1, 0)

Impulse:

Step block and derivative block

(0.01, 0, 1, 0)

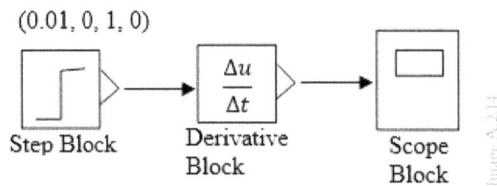

Step Block   Derivative   Scope
               Block   Block

Example

step   sum   gain   zero-pole   Transfer Function   scope

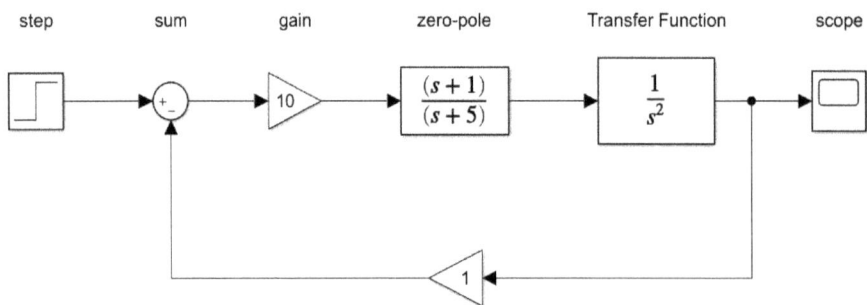

$$\frac{(s+1)}{(s+5)}$$

$$\frac{1}{s^2}$$

$$10$$

$$1$$

In the MATLAB window, type "Simulink" or choose Simulink from the top menu bar. Choose the feedback controller. First, create a working area by selecting "NEW" under File menu. Under "NEW," choose "Blank model."

A working window named "untitled1" will appear. Under the "Tools" menu, choose "Library Browser."

Now you can start building a model by dragging the blocks from the library, dropping them into the working window, and connecting them. To connect two blocks, click on the output port of one block and drag the mouse to the input port of another block. In the example, there are six blocks in a forward loop and one block in a feedback loop. Drag STEP, SUM, GAIN, ZERO-POLE, TRANSFER FCN, and SCOPE blocks one at a time into the working window. Connect all the blocks as shown. Drag another gain block and double-click on it. From the drop-down options, choose "FLIP IT" to flip the block and place it in the feedback loop.

Open the step function block by double-clicking on it. Enter "0" for the initial starting time. Enter the initial value and final value as "0" and "1," respectively.

In the summing block, double-click to open it and change the feedback sign from 'plus (+)' to 'minus (−)'.

In the gain block, open the block and enter the value 10.

Open the zero-pole block and enter the zero-pole vectors as

Image A.2.16

Open the transfer function block and enter the numerator and denominator as a vector of polynomial coefficients.

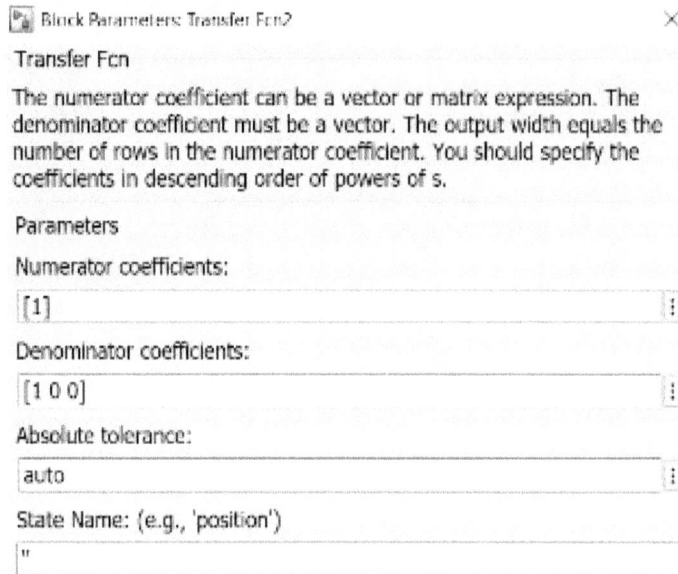

Block Parameters: Transfer Fcn2 ✕

Transfer Fcn

The numerator coefficient can be a vector or matrix expression. The denominator coefficient must be a vector. The output width equals the number of rows in the numerator coefficient. You should specify the coefficients in descending order of powers of s.

Parameters

Numerator coefficients:

[1]

Denominator coefficients:

[1 0 0]

Absolute tolerance:

auto

State Name: (e.g., 'position')

''

Open the feedback gain loop and enter "1."

Running Simulation

From the simulation menu, select "START," or select "RUN" on the menu bar. To view the result, open the SCOPE block by double-clicking on it. The output will be displayed in a graph.

Save the Model

It is important to save the model by selecting "Save as" and giving the file a name. Then, you can come back to the model in the future by choosing "Open" in the Files menu. This will show your saved files. It is easy to change the gain or transfer function by opening the respective block and entering the new values. It helps study the output when the parameters change without building another model or writing a program code.

CREDITS

Appendix B

Answers to Practice Problems

Chapter 1

1. 1-c, 2-b, 3-d, 4-b, 5-a, 6-d, 7-d, 8-b, 9-c, 10-b

2. Open loop with input—time, output—bread color

3. Closed loop with input—time, output—bread color and a feedback sensor for bread color

4. Bimetallic strip, gas-filled bellows, expanding wax pellets, electrical thermocouples, electronic thermistors, and semiconductors

5. Body (skin), mouth (tongue), eyes, nose, ears

6. Speed—throttle, gas pedal, speed sensor; direction—GPS, steering, direction sensor

7. $Y(s)/R(s) = s/(s + a)$

8. $F(1/K) = x$; F-input, x-output

9. $T(s) = (s + 4)/(s^2 + 4s + 1)$

10. $T(s) = (s + 6)/[s(s^2 + 5s + 4)]$

Chapter 2

1. 1-d, 2-b, 3-d, 4-a, 5-c, 6-b, 7-a, 8-a, 9-a, 10-b

2. $[M] = \begin{bmatrix} m_1 & 0 \\ 0 & m_2 \end{bmatrix}$; $[K] = \begin{bmatrix} k_1 + k_2 & -k_2 \\ -k_2 & k_2 \end{bmatrix}$

3. $[M] = \begin{bmatrix} m_1 & 0 & 0 \\ 0 & m_2 & 0 \\ 0 & 0 & m_3 \end{bmatrix}$; $[K] = \begin{bmatrix} k_1 + k_2 & -k_2 & 0 \\ -k_2 & k_2 + k_3 & -k_3 \\ 0 & -k_3 & k_3 + k_4 \end{bmatrix}$

4. $[M] = \begin{bmatrix} m_1 & 0 & 0 \\ 0 & m_2 & 0 \\ 0 & 0 & 0 \end{bmatrix}$; $[C] = \begin{bmatrix} c_1 & 0 & 0 \\ 0 & c_2 & -c_2 \\ 0 & -c_2 & c_2 \end{bmatrix}$; $[K] = \begin{bmatrix} k_1 & -k_1 & 0 \\ -k_1 & k_1 + k_2 & -k_2 \\ 0 & -k_2 & k_2 \end{bmatrix}$

5. $G(s) = \dfrac{1}{(Det)} \begin{bmatrix} (2s^2 + 5s + 2) & (2s + 1) \\ (2s + 1) & (s^2 + 3s + 2) \end{bmatrix}$; $(Det) = 2s^4 + 11s^3 + 17s^2 + 12s + 2$

6. $[M] = \begin{bmatrix} 2 & 0 & 0 \\ 0 & 3 & 0 \\ 0 & 0 & 4 \end{bmatrix}$; $[C] = \begin{bmatrix} 5 & -5 & 0 \\ -5 & 11 & -6 \\ 0 & -6 & 6 \end{bmatrix}$; $[K] = \begin{bmatrix} 40 & -10 & -25 \\ -10 & 25 & -15 \\ -25 & -15 & 60 \end{bmatrix}$

7. $[Z(s)] = s^2[M] + s[C] + [K]$; $[G(s)] = [Z(s)]^{-1}$; use MATLAB

8. Poles: $-0.9 \pm j2.86$; zero: -2.5

9. Poles: $-1.5 \pm j4.77$; zero: -3; $|s| = 5$; $\theta = -72.54 \, deg$

10. (a) $y(t) = [-4e^{-t} + e^{-3t} + 3]$; $y(\infty) = 3$

11. (a) $y(t) = \dfrac{15}{4}[e^{-t} - 2e^{-3t} + e^{-5t}]$; (b) $y(t) = \dfrac{1}{60}[225e^{-t} - 50e^{-3t} + 9e^{-5t} + 120t - 184]$

12. MATLAB exercise

13. Simulink exercise

Chapter 3

1. 1-d, 2-b, 3-a, 4-d, 5-b, 6-c, 7-d, 8-d, 9-b, 10-d

2. $T(s) = \dfrac{1}{s^2}(s^2 + s - 2)$

3.

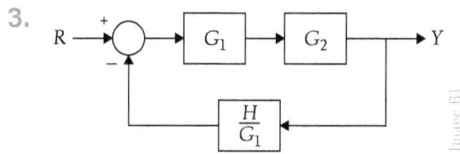

4. $T(s) = \left[\dfrac{G_1}{1 + G_1 G_2}\right] G_2$

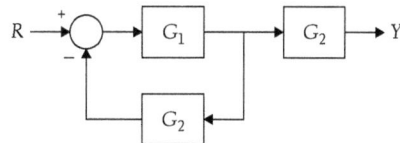

5. $T(s) = [KG_1 G_2]/[s(1 + G_1 H_3 + G_1 G_2 H_2 + G_1 G_2 H_1) + KG_1 G_2]$

6. $T(s) = \dfrac{4(s^6 + 2s^5 + s^4 + 14s^3 + 28s^2 + 14s)}{(s^{10} + 3s^9 - 45s^8 - 125s^7 - 200s^6 - 1177s^5 - 2344s^4 - 3485s^3 - 7668s^2 - 5598s - 1400)}$

7. $T(s) = [k_1 k_2]/[s^2 + s(k_1 + k_1 k_2 + k_2 k_3) + k_1 k_2 k_3]$

8. $T(s) = \dfrac{(G_1 G_2 G_3 G_4 G_5 G_6)}{[1 + (G_1 + G_2 + G_4 + G_6) + (G_1 G_4 + G_1 G_6 + G_2 G_4 + G_2 G_6 + G_4 G_6) + (G_1 G_4 G_6 + G_2 G_4 G_6)]}$

9.

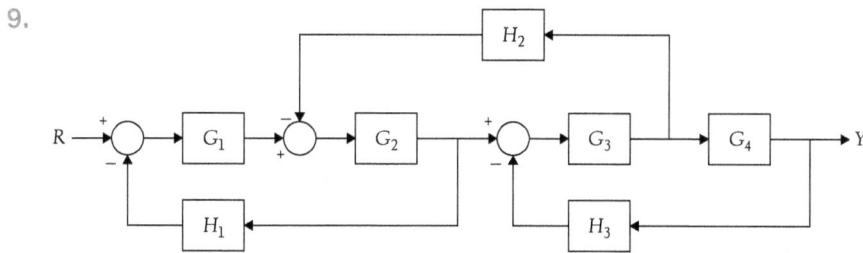

10. $T(s) = \dfrac{(G_1 G_2 G_3 G_4 G_5)(1 + G_8 + G_{10} + G_8 G_{10}) + (G_6 G_7 G_8 G_9 G_{10})(1 + G_3 + G_5 + G_3 G_5)}{[1 + (G_3 + G_5 + G_8 + G_{10}) + (G_3 G_5 + G_3 G_{10} + G_3 G_8 + G_5 G_8 + G_8 G_{10}) + (G_3 G_5 G_8 + G_3 G_8 G_{10})]}$

11. Simulink exercise

Chapter 4

1. 1-d, 2-c, 3-a, 4-b, 5-d, 6-b, 7-b, 8-a, 9-d, 10-c

2. $E(s) = (\tau s + 1)/(\tau s + 11)$; (b) $s_\tau^T = \dfrac{-\tau s}{(\tau s + 11)}$

3. At $N = 0$, $Y/R = \dfrac{20(s + 10)}{s^2 + 30s + 220}$; (b) at $R = 0$, $Y/R = \dfrac{-20}{s^2 + 30s + 220}$

4. $Y = [G/(1 + GH)]\,(R) + [1/(1 + GH)]\,(T_d)$

   (a) $Y/R = K(s^2 + 5s + 6)/(s^3 + 15s^2 + 56s + 60 + 14K)$

   (b) $e_{ss} = (60 + 8K)/(60 + 14K)$;   (d) $y(\infty) = (60)/(60 + 14K)$

5. $Y = [GG_c/(1 + GG_c)]\,(R) + [G/(1 + GG_c)]\,(T_d)$

   (a) $[(0.1s + 1)(s^2 + 25s + 200)]/[(0.1s + 1)(s^2 + 25s + 200) + 200K]$

   (b) $[200(0.1s + 1)]/[(0.1s + 1)(s^2 + 25s + 200) + 200K]$

6. (a) $Y = [GG_c/(1+GG_c)]\,(R) + [G/(1 + GG_c)]\,(T_d)$

   (b) At $T_d = 0$, $T = Kb/(s + Kb + 2)$; (c) $s_b^T = (s + 2)/(s + Kb + 2)$

7. $e_{ss} = 0.04$

8. $K = 498$

9. $K = 5$

10. $K_1 = 0.9$; $\tau_c = 0.015$

Chapter 5

1. 1-a, 2-d, 3-c, 4-b, 5-c, 6-d, 7-a, 8-b, 9-d, 10-d

2. (a) At $\delta = 0.01$, $(T_s/\tau) = 4.6$; At $\delta = 0.1$, $(T_s/\tau) = 2.3$

   (b) $\tau = 0.21$; $T_s = 0.53$ sec

3. $T_s = 2$ sec; $PO = 25.4\%$; $T_p = 0.686$ sec; $T_r = 0.3$ sec; $M_{pt} = 1.254$

4. $\tau = 0.25$; $T_p = 0.4$ sec; $T_r = 0.18$ sec; $M_{pt} = 1.2$

5. (a) P.O. $= 4.3\%$, $T_s = 4\sqrt{2}\,/\sqrt{K}$;   (b) $K > 32$

6. (a) $e_{ss} = 0.97$;   (b) $G_p = 30$

7. $K \geq 4$

8. $K = 7.51$; $p = 4$

9. $K = 3$

10. $K = 16$; $K_1 = 5.36$; $e_{ss} = 0.335$; $s_{K_1}^T = (-5.36s)/(s^2 + 5.36s + 16)$

Chapter 6

1. 1-d, 2-b, 3-a, 4-a, 5-a, 6-d, 7-b, 8-a, 9-c, 10-d

2. (a) Stable; (b) unstable; (c) unstable

3. Unstable

4. Stable for $0 < K < 6.875$

5. Stable for $0 < K < 20$

6. Stable

7. For stability, $0 < K < 2$

8. Unstable

9. Marginally stable

10. For stability, $20 < K < 108$

Chapter 7

1. 1-a, 2-a, 3-b, 4-d, 5-b, 6-d, 7-c, 8-b, 9-d, 10-d

2. (a) $P(s) = (s + 5)/(s^2 + 4s + 3)$

   (b) Breakaway point $= -2.17$, $K = 0.343$; break-in point $= -7.83$, $K = 11.66$

3. (a) Break-in point $= -2.414$, departure angle $= \pm 225°$

4. Asymptote: center $= -2/3$; angle $= 60°, 180°, 300°$

   Angle of departure $= \pm 333°$

   Crossover point: asymptote $= \pm 1.155j$; loci $= \pm 2.236j$

5. Breakaway point $= -0.88$; $K = 4.06$

   Asymptote: center $= -2.33$; angle $= 60°, 180°, 300°$

   Crossover point: asymptote $= \pm 4.04j$; loci $= \pm 3.16j$

6. (a) $G_c = (1/s)(11.25s + 5.4)$; (b) $G_c = 15 + 4.69s$

7. $G_c = (1/s)(36.86s^2 + 280.8s + 534.86)$

8. $G_c = (1/s)(341.72s^2 + 628.76s + 131.22)$

9. $K_{P1} = 9.2$; $K_{P2} = 12.25$; $K_D = 2.45$; $K_I = 4.9$

10. $K_U = 160$; $T_U = 1.57$ sec

11. MATLAB exercise

12. MATLAB exercise

13. Simulink exercise

14. Simulink exercise

Chapter 8

1. 1-a, 2-b, 3-a, 4-c, 5-b, 6-a, 7-d, 8-c, 9-d, 10-a

2.
ω	10	20	30
dB	127	146	159
Φ (deg)	3.06	4.57	5.13

3. $dB = -138.9$; $\Phi = -17.3$ rad (-271 deg)

4.
ω	0	0.5	1	2	4	10	100
dB	0	−0.53	−1.94	−6.02	−13.98	−28	−68
Φ (deg)	0	−28.07	−53.13	−90	−126.87	−157	−178

5. $(\omega)_B = 7.27$ rad/s

6. $(\omega)_B = 5$ rad/s; $(\Phi)B = -186.34$ deg (-3.25 rad)

7.
ω	0	1	10	100	∞
$r = |L|$	62.5	55.7	9.4	0.15	0
Φ (deg)	0	−31.3	−118.5	−172	−180

8. Stable for $0 < K < 27$; marginally stable for $K = 27$; unstable for $K \geq 27$

9. PM $= -52.560$, GM $= 1.94$ dB

10. $(\omega)_n = 4$; $\zeta = 0.5$; $(\omega)_c = 3.145$; PM $= 51.8$ deg

Chapter 9

1. 1-c, 2-a, 3-b, 4-c, 5-b, 6-c, 7-b, 8-a, 9-a, 10-b

2. $s = -4.45 \pm j3.57$; $Gc = [90.5(s + 0.48)]/[s + 43.5]$

3. $s = -2.7 \pm j3.55$; $Gc = [13.93(s + 0.62)]/[s + 8.66]$

4. (a) Compensator is required

   (b) $Gc = [130.58(s + 0.4)]/[s + 49]$; $s = -8 \pm j6$; compensator pole: $s = |-49| > |-8|$

5. $K_P = 14$, $K_I = 64$, $G_c = (2/s)[7s + 32]$

6. $K_P = 28$, $K_I = 32.5$, $G_p = 1.16/(s + 1.16)$; $G_c = (1/s)[28s + 32.5]$;
   $T(s) = (32.5)/[s^2 + 8\,s + 32.5]$

7. $G_p = 5/(s + 5)$; $T(s) = (10)/[s^2 + 5s + 10]$

8. $G_p = 8/(s + 8)$; $T(s) = (80)/[s^2 + 6s + 80]$

9. $G_c = [5.1(s + 0.1)]/[s + 0.01]$

10. $G_c = [23.2(s + 0.15)]/[s + 0.02]$

Chapter 10

1. 1-c, 2-a, 3-d, 4-a, 5-c, 6-d, 7-d, 8-c, 9-a, 10-c

2. $[A] = \begin{bmatrix} 0 & 1 & 0 \\ 0 & 0 & 1 \\ -2 & -5 & -4 \end{bmatrix}$; $\{B\} = \begin{Bmatrix} 0 \\ 0 \\ 1 \end{Bmatrix}$; $[C] = [5\ 6\ 2]$

3. Block diagram:

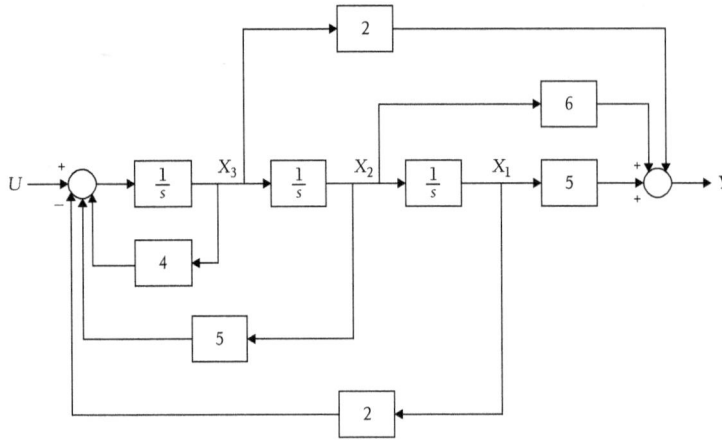

Signal Flow diagram:

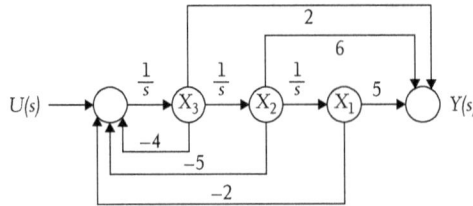

4. $y(t) = (3.5e^{-2t} + 7e^{-t} + 5te^{-t} + 2.5)$

5. $K = 2$

6. $k = 2$

7. $k_1 = 13, k_2 = 4$

8. Controllable; not observable

9. Controllable; observable

10. $[K_1 \quad K_2] = [-3 \quad 9]$

www.ingramcontent.com/pod-product-compliance
Lightning Source LLC
Chambersburg PA
CBHW061351210326
41598CB00035B/5948